AWKLT3 PRESS

Beyond the Microphone

Restoring Truth. Reviving Hope. Rebuilding America.

LT of *And We Know*

Beyond the Microphone
SEMPER FI

LT's Life Story and How *And We Know* Was Born

And we know that all things work together for good to them that love God, to them that are the called according to His purpose. —Romans 8:28

Beyond The Microphone
Semper Fi

By LT – Founder and host of *And We Know*®

AWKLT3 Press
Copyright ©2023. All Rights Reserved.

Under International Copyright Law, no part of this book may be reproduced, stored, or transmitted in any form or by any means without written permission from the author/publisher.

All Scripture quotations and references (unless otherwise marked) are taken from the King James Version (KJV) of the Holy Bible.

ISBN Number: 979-8-9881202-0-9

Editor and Book Design: Gene Fedele
Cover Design: Emilie Westby
Back Cover Photo: ©And We Know

Printed in the USA by AWKLT3 Press

AndWeKnow.com

CONTENTS

Dedicatory 7

Author's Introduction 9

Chapter 1: Here's Your Good Sign 11

Chapter 2: A Good Foundation 17

Chapter 3: The Good Drive 29

Chapter 4: A Good Amen 35

Chapter 5: Good Faith to Move Mountains 43

Chapter 6: The Taj Mahal Was Good! 49

Chapter 7: The Good Walk 61

Chapter 8: My Three Good Kings 71

Chapter 9: Cursing, Gothics, Satanists = Good? 87

Chapter 10: God's Good Hand in Broadcasting 99

Chapter 11: Ain't God Good? 111

Chapter 12: Walk on Good Water 129

Chapter 13: The Good Daily Bread 141

Chapter 14: Vengeance is Only Good from the Lord .. 153

Chapter 15: LT, the Good Start! 161

Epilogue: LT's Good Testimony! 167

DEDICATORY

As I was working on this book, it became quite clear to me that my life wouldn't be where it is today without my wife, Mona, and our five amazing children. They have all had a huge impact on my life, and they have had to deal with all of my ups and downs, adventures, dreams (some that went nowhere), long lectures on why the world is so messed up, super long sermons in the car, 80's music and more.

Mona, since the day I first met her, has always seen something in me and encouraged me to "Go for it." When I am feeling down, she has a way of picking me up. Her smile and laughter throughout our life has been the greatest encouragement any husband could ask for. She has always looked for ways to keep our marriage strong, and also has a way of making sure that other people and friends know she is proud of being married to me, when there are certain days she probably has been challenged to believe it. Can I get an *Amen*?

I want to thank all of my children, Brandon, Anthony, Rafael, Noah and Maria for believing in me. You each display a love for me and others that has been so encouraging. Your mom and I are so proud of you, and we thank the Lord for your amazing gifts.

Of course, none of this would have been possible without the *Amazing Grace* and *Love* of our Lord and Savior Jesus Christ. Without Him we are nothing.

AUTHOR'S INTRODUCTION

Well, the time has finally come to tell the story of LT and the birth of *And We Know*. Sitting on a plane over the Pacific, my thoughts wander to how this will all play out. What do folks want to know and how it will all be received? What impact can I make in people's lives and across the earth?

I am reminded to simply pray and remember the purpose of writing this book. My hope is that it will challenge everyone to not only read the Word of God (The Holy Bible), but to put it into practice in every facet of their lives.

Are you experiencing a storm on the day when you really needed the sun to come out? Are you working with someone who seems to have all the signs of "evil?" Is there doubt that you should live your life to the fullest for Christ? Have you found yourself doubting your direction in life? Are you living in fear because of a job loss or uncertain future?

We have all faced difficult circumstances at some point in our life. Do we call it good? Often doubts have crept in. What will life be like when I leave home for the first time? Will the difficult circumstances work out for good? Can I make it as a parent in this crazy world? Will we find a job when God removes the one we depend on? How can we trust our Heavenly Father as we endure our daughter's life and death battle?

Yet, there's more to this story for us to uncover. I hope and pray for divine inspiration for every reader to be encouraged and lifted up through life's difficulties and trials, and to turn to God and His Holy Word for direction, hope and faith to persevere.

    Our Lord's
        Humble Servant,
           LT.

From my study at *And We Know*
May, 2023

Chapter One

Here's Your "Good" Sign!

And we know all things work together for good to them that love God, to them who are the called according to His purpose.
[Romans 8:28]

"Who here is joining the Marine Corps?" asks the man in some type of military uniform at the Atlanta airport. As about 40 men and women looked around the dimly lit room, my hand slowly went up along with one other guy.

"Your training will be 12 weeks long," he said as the room filled with laughter and whispered comments like "Those guys are crazy" or "thank God I didn't sign with them."

As the military folks continued to ask where all the other future young men and women were going, my mind turned to thoughts at that moment that I might "never see my family again." Quite sobering to say the least.

As we made our way into the airport hallway and formed a single-file line, I quickly ran to a photo booth, deposited a dollar, and four small pictures showed me running my hands through the long light-brown hair that took me so long to grow. Adding to that pain and memory were the four women who walked down the line as we stood at attention, letting each one of us know they were thankful for our "future service." For some reason, they felt obligated to stop next to me with a deep look of sorrow.

"Oh man…would you look at that," one lady told the others as she placed her hands on my hair. "All that beautiful hair is going to be gone for the rest of his life. That is so sad."

Tears started flowing (mine, of course).

We eventually split up to board our plane destined for Marine Corps recruit training in the wonderful low country town of Beaufort, South Carolina. Upon arrival, we joined 80 more young men who we met up with in a room at the airport. We eventually ended up on a bus at midnight for a journey every Parris Island Marine remembers.

I traveled on many buses throughout my life—school buses, youth-trip buses, church buses on Sunday, high school baseball team buses, Greyhound buses, to name a few; and there was one thing they all had in common—joy, laughter, fun and games.

We future Marines boarded a bus in Charleston, South Caroline, filled with excitement. We shared our stories of how the other branches of the military treated us in other locations as we made our way to this point in the adventure in our life. Many showed pictures of their brothers and sisters, girlfriends, moms and dads and more. The chatter was nonstop until we made the infamous "right turn" and a military policeman with a chiseled face and a pistol on his side waved the bus slowly through a gate which said, "U.S. Marine Corps Recruit Depot–Parris Island, SC."

Complete silence. The bus made its way over a dark bridge. Look to the right darkness. Look to the left more darkness, due to the water surrounding the island.

Eventually, large trees with flowing Spanish moss appeared, but then there is more darkness. Thoughts ran deep on this short trek from the gate to the famous yellow footprints[1]. Mostly questioning the reality of the moment. "I have to remember this route for my escape," or "what was I thinking?"

"Get off my bus!" screams the drill instructor after giving a long, stern speech welcoming us to the island. As I was thinking about how much I was already missing my home, parents and the familiar, he proceeded to let us know that he would be our mother and father for what we perceived was "the rest of our lives."

---

1. **Yellow Footprints** are painted on the ground at Marine bases to direct newly arriving recruits where to stand in order to receive instructions.

All Things Lead to Good

Now, let me pause here to let you know that as you are reading this journey it leads to a point of goodness. That goodness will somehow make its way like a thread sewing a beautiful tapestry together. So, let's continue.

After listening to a constant barrage of screaming, changing uniforms, making a 30-second call home, having all our hair shaved off…I felt a sense of accomplishment, having survived the first 24 hours. At least that's what it felt like, and if I remember this correctly, they forced us to stay awake for a couple of days straight…I guess as an initiation of some sort.

"Next!" screams a Marine behind a small window.

"Hey, you, get your butt over here now!" pointing to me.

The Marine sergeant I faced waded through my very thin Service Record Book (SRB). He looked at the job that the recruiter promised me a year earlier and comments, "Aerial Navigator… wow, your recruiter got you didn't he. This is only for warrant officers. We are going to change your job. Get out of my face recruit!"

Wait a minute. I wanted to be assigned to planes. "Why would God do this to me?" I thought. The youth group just prayed for me last week. The pastor of the church said that he felt that God was going to use me for something huge. This Marine just gave me the worst news ever…all after being put in this prison called Parris Island with no way out, no hair and a room filled with the stench resembling football players who finished their big game last night but never changed out of their sweaty uniforms.

Serving "In the Air"

I then recalled the day my Marine Corps recruiter talked to me one year earlier about my career preferences.

"LT, your ASVAB score is very high," said the Marine Gunnery Sergeant. "You can have any job you want in this book.

The ASVAB is the Armed Services Vocational Aptitude Battery that measures your abilities. It's a multiple choice test that can somehow figure out what job you would be proficient in. It's as if someone hands you a basketball and tells you to shoot a basket and you make it 40 out of 50 times…you might be great for basketball.

"I want to be in the air sir," I responded.

"Ok, well this job (Aerial Navigator) would have you in a C-130 all the time. Let me make a call to see if it is available," the recruiter said.

After a brief conversation with someone on the other line, he said "The job is available, but you only have one week to grab it or this MOS[2] will be gone…aerial navigator."

Now, he said this on August 26, 1987. I suspect he needed another person to sign up to meet his 2-person monthly quota.

After pouring through encyclopedias and the library on aerial navigation and with very little sleep, I was in the recruiter's office signing my life away into the Delayed Entry Program as a senior in high school.

I signed up on August 28, 1987 and went to bootcamp one year later Aug. 28, 1988.

Hmm…8/28/88 = Romans 8:28. AND WE KNOW.

As Bill Engvall or Jeff Foxworthy would say, "Here's your sign."

No matter how much I avoided that recruiter or what circumstances led me to that office, I believe the Lord wanted me to remember He was in charge and someday those numbers would remind me that He always works out "all things for good."

He knew that 30 years later, in August of 2018, I would be told by my teenage son Rafael, "Dad you need to record yourself talking about all of these things about President Trump and the Deep State. I don't understand it, but you make me love listening to you."

Life in Boot Camp

After hearing that the job the recruiter promised to me was going to change, it deeply affected me and I wanted to escape. We were finally taken to a barracks room with bunkbeds after our lives were ripped apart in under 48 hours. The drill instructor belted out orders in an effort to control every motion we made.

---

2 Military Occupational Specialty

"This is your home now recruits," said this robotic, slim fit man with a tired voice, as if he had been through hell with us. I would later discover when stationed at Parris Island as a Marine just how tired these Drill Instructors really were.

"Take out that toothbrush, now."

"Aye, aye sir," we screamed out in unison. "Louder!"

"Aye, aye sir," managing to get enough strength to belt it out.

"You have one minute get those teeth cleaned. 60, 59, 58, 57."

We ran to the head (restroom) and when I looked in the mirror I jumped! It was the first time an "alien" stared at me. My new white marble head was startling.

This was one of many "depressing" moments throughout this journey in "prison." Watching the largest young men cry on the floor that first night, some saying, "I am never going to see my family again." Now, I know this is a hard scene to imagine, but that moment in time is still fresh in my mind and filled with horror.

This was the first night of freedom away from all the shouting. It was the first night I was able to actually think without someone thinking for me. I thought, "Why did you bring me here God? I want to go home."

I can only imagine how Daniel felt in the Bible. In the first chapter of the book of Daniel, it appears that Daniel's life was turned upside down as he was taken from his family and brought to the king. But later on we read that God blessed him with the gift of wisdom, so much so that the king wanted his talent in his court. Did Daniel question God? Did he complain about his situation? Or did he exercise faith in God and surrender to Him realizing that his life would be used someday for a good and godly purpose?

That night I realized that though our lives were being directed to serve an earthly king in government, what an opportunity was before us to see our lives used for our Heavenly Father's purposes?

And be not conformed to this world: but be ye transformed by the renewing of your mind, that ye may prove what is that good and acceptable, and perfect, will of God.—Romans 12:2

CHAPTER TWO

A Good Foundation

Trust in the LORD with all your heart, and do not lean on your own understanding. In all your ways acknowledge him, and he will make straight your paths.
[Proverbs 3:5-6]

I was born in the small town of Socorro, NM about an hour's drive South of Albuquerque. My father's side of the family were almost all Roman Catholic, and my mother's side of the family attended the local Baptist church.

Not too long after my birth, my dad decided to join the U.S. Marine Corps, that meant we moved around a lot; however, we made many trips to Socorro throughout my early years prior to my Marine Corps life.

In many ways, when considering my family dynamics, I would say that "spoiled" is an understatement when it comes to describing both sides of each family. I spent a lot of time with both sets of grandparents and great grandparents. The influence that they had on my life contributed greatly towards building a solid foundation for who I am today.

My dad's side of the family was quite wealthy. They owned a gas station, hotel and apartments. My mom's side of the family didn't have as much on the monetary side, but their outpouring of love was extraordinary.

During those formative years of spending time with different family members, no matter who I was spending time with, one non-negotiable that was implanted into our minds was that Sunday is the Lord's Day—the day we attended church. As a result, I had the opportunity to see both the Roman Catholic and Baptist forms of worship and church life.

My great grandmother from my father's side was so precious. She would often invite me over for pancakes with her little gray poodle "Hootie" always by her side. Her perfume was very strong, and I miss hearing her say in her wonderful Spanish accent, "Aye Miho, you are so good looking. Come in, I made some pancakes for you."

My dad's mother and father, Mary and Lawrence, loved their children, grandchildren and great grandchildren. Almost every time we paid a visit, their children seemed to all show up, bearing gifts and with a keen interest in checking the kitchen to snatch a plate of the latest chili and tortilla meal on the stove. Everyone would share how their latest life stories, how business was moving along and even bring up the latest news on the TV. I fondly remember all the smiles and hugs liberally shared in our family!

Throughout the years, I had so many memorable opportunities to spend time with them and I remember, with much affection, the numerous excursions we took together to Disneyland, Sea World, Knottsberry Farm, camping in Colorado and so much more.

My mother's grandma was just as sweet. I loved sitting and rubbing her feet when she wasn't feeling good, and she would tell me stories about her two favorite "J's" in life…Jesus and John Wayne. Her husband, who comes from an Irish background as a Lloyd, grew up a deacon in his church and loved to preach the Word of God. Each time we visited their house in Oklahoma, we knew we were in the presence of a holy man empowered by the Holy Spirit.

My mother's mom, Effie, and her husband R.C., would fill their home with all my aunts and their husbands and our cousins whenever we stopped by for the holidays or summer. We would enjoy playing dice games, exploring the property for memorabilia, singing, and always praying together before dinner time. They also loved to hug and cry and talk about what Jesus was doing in their lives. One thing that really made an impression on me throughout the years was sitting in my grandma Effie's Sunday School class, listening to her teach those Bible precious stories with a deep passion and love for our Savior.

Around the age of eighth, I also experienced great love from another family as we gained a stepdad, also a Marine, who had very strong Lutheran upbringing in North Dakota. His prayers were very powerful and was also a part of our nightly pre-dinner traditions.

His parents had a love for gardening, and they would each write me very long, hand-written letters throughout my life. They were filled with the love of Christ and their awe for His creation and how the earth was filled with so many gifts—even providing for their canned jelly and their frozen winters.

My positive family experiences and godly examples have been foundational towards building my relationship with my Heavenly Father. I was able to see and experience love from family that seems to be missing from many folks I've met throughout the years.

While family was important in building this strong foundation for my future, one of the most memorable experiences in strengthening my Christian walk occurred in Alliance Christian Academy in Albany, GA, during my middle school years. My mother worked with the school to secure an affordable tuition for her children and she also ensured the Marine Corps base provided a bus to get us to that school. She was determined to find us a decent education after our public-school English teacher sent a letter home with four misspellings in the first paragraph.

We memorized Bible verses, had "Sword" drills (competitions used to see who would be the first to find Bible verses), were visited by missionaries from around the world, and had to perform weekly book reports by standing in front of the classroom with only a few note cards.

This gave me a love for competition. At the end of the second year, I received the award for the highest grade in almost every subject we studied and participated in school.

My drive to win carried over into other areas of life, including Boy Scouts and Civil Air Patrol. These also contributed to a strong foundation for my life. We were always challenged to get the next merit badge, win the Pinewood Derby race or fishing competition, or simply make the best kites in the world (which wasn't so simple, by the way).

My drive to excel translated into sports as well. I grew up with a love for baseball and soccer. Although my number one goal was winning, I was always felt a sense of achievement upon making the All-Star teams.

It seemed I was always out to win. I think this might have been because I was always the smallest guy on the field or in school, standing at only 5 feet 4 inches tall.

My parents were no strangers to competition, which I, no doubt, picked up throughout my upbringing. My mother was always participating and winning ceramics contests, getting the lead singing role in the choir and making the best cakes in town. My father was the best golfer in the U.S. Marine Corps in 1976 and was also the strongest in the bench-press competitions. My stepdad was able to get one of the highest civil service positions in the calibration department aboard MCLB Albany, Ga., He was also determined that we would win the "Yard of the Month" on the military base, and he showed me neat tricks on how to make the perfect cars for the pinewood derby competitions (which we won or were a finalist several times).

Added to this was my mom becoming a nurse while taking care of six children, my dad becoming one of the top CPAs in Navy Federal, and my stepdad getting the most coveted position on the Marine Corps Base as a civilian.

You see, God ensured the right examples were always placed in front of me. We didn't grow up in perfection, of course. Nobody does. But through our imperfections God has a way of bringing it all together to work for GOOD.

Surrendering My Life to Jesus
Although my upbringing was strongly rooted in Jesus Christ and understanding that He was the Savior of the world, the key moment where I officially gave me life to Christ was at the age

of 16 during a youth meeting in the church. After the message, my mind was filled with awe and wonder at the saving grace of Jesus Christ. It was a very emotional night. I walked forward to pray and simply knew at that moment that Jesus was my Savior. A few weeks later that same youth pastor who gave the message pulled me into his office and said, "this might seem strange to you now, but I sense that God is going to use you for something very big someday for His Kingdom. I hope you remember me telling you this."

He had no idea then that *And We Know* would be what he was talking about. Before I left for boot camp in 1988, the youth group stopped by our house to pray for me and they all agreed with one of my friends that night that "God is going to use you to preach the gospel to thousands as a chaplain or something… we aren't quite sure, but you are going to do big things for Christ."

The Three "M" Principle, Add One "M" for GOOD Measure

In building a "GOOD" foundation, the Lord had to get some things straight in my life. It would take many years to have those predictions from the youth pastor and the youth group to come true. He had to really work hard on me.

Our beloved Pastor Abbott told us years later that God had a three "M" principle designed to get you straightened out.

1. MASTER – Who is the master of your life? Some have video games, some let addictions lead them, but we should have God as our Master.

2. MIND – How will you use your mind. If you get your Master correct, you can start reading the Word of God to transform your mind.

3. MATE – "If you have the right master who can transform your mind, He alone can ensure you have the right wife," pastor Abbott said with a deep voice.

Well, God threw in an extra "M" for me alone. Her name is Mona. We are in our 31st year of marriage as of 2023. God saw I was trying to do the reverse by finding a "mate" first.

You see, I had a large family with many sisters. Being a part of a large family instilled a drive for romance and a longing for my own family as I grew up. Each time I would go on a date, I would fall in love and was determined that "She is the one." I could honestly say that this started at the age of ten as I was riding the military bus to the pool each day. To one poor young girl, I was already proposing and giving my turquoise ring away. My soul was crushed the day she moved out of town due to a military move.

Throughout high school and the first few years in the Corps, I longed to be in a relationship. Looking back, now that I am a big "conspiracy" guy, the influence of *Little House on the Prairie* and Hollywood movies probably made me feel that was important at the time.

Almost every girl I dated broke it off, sending me into a deep depression each time.

God finally put a stop to this search for "Treasure" in 1991 while stationed in Okinawa, Japan.

Finding My Soul Mate
I met Mona in Okinawa as a young corporal in the Corps on Marine Corps Air Station Futenma in Okinawa. She was volunteering to help her friend at the local store across the street from our barracks.

The first time I noticed her was at the doorway of our barracks. She was wearing a fine woman's suit, had perfect straight hair that went to her shoulders and there was ¼ inch blonde strip across her brow. Her legs were very nice. That is something that you never forget as a young man.

What really caught my attention was the fact that many of the Marines were trying to ask her out on a date throughout that first week she was there…and they all struck out.

"I think she is married," one Marine replied appearing depressed that Mona wouldn't give him the time of day.

Others speculated that she was dating someone or was simply "mean."

Now that was very attractive to me. I thought, "I need to figure out how to meet this 'tough' girl."

So, I came up with a plan, a Marine Corps plan that many Marines know as BAMCIS. Begin the Plan, Arrange Reconnaissance, Make Reconnaissance, Complete the Mission, Issue Orders and Supervise.

Each barracks had a "momma-san" who would do all our laundry for $20 every two weeks. My momma-san was special, and she knew me for more than 2 years. I asked her for help.

"Momma-san, can you help me meet the girl across the street?" I asked in humility.

"What you want me to say?" she asked in broken English.

"Tell her I am a nice guy," I said.

"You not nice guy," she said in her broken English as she slapped me on the shoulder and continued, "Come on, I help you."

As she talked to my future wife, I was around the corner hiding and listening to the conversation. That tiny five-foot tall, 100-pound momma-san reached around the corner and pulled me into the room.

"Here he is," she said to Mona as she smiled and laughed.

"I wanted to say 'hi' and let you know that I think you are pretty," I said.

"I'm not pretty," she said, as she turned different shades of red.

"Nice to meet you," I said, as I left feeling embarrassed and thinking that I had failed.

Each day, I dropped by to simply say "hello." She had just returned from the United States where she was an exchange student at Old Dominion University, and she attended a small Presbyterian church with her aunt for a year. Her English was very good. Her Okinawan parents, who are so amazing, told her to stay away from Marines which explains why she was so tough. Through that toughness was a very happy, always smiling young lady with a great work ethic.

One day she really caught my attention when I noticed she was using the barracks washing machine and dryer to take care of her parent's clothes. She would end each day of work by folding clothes and putting the basket in the back of her car. They had their own washer at home, but she enjoyed taking care of her parents and said it was faster on the military base.

Another thing that impressed me was when I went to visit her at her permanent work office on Camp Foster on Okinawa, she was always talking to full-bird Marine Colonels who loved talking to her as she has an attractive gift of making others feel special and important.

After a few months of dating, we attended a small party for our unit at one of our Master Sergeant's homes on Camp Kinser. She thought I was "very important" as all my Marine friends stood at attention as I walked her to the back yard. The Master Sergeant was married to a wonderful lady from the Philippines, and they all really enjoyed Mona. She has a way with people that is so wonderful and inviting.

The next day at work, the Master Sergeant pulled me into his office. Back in those days, no one was allowed in the "Top's" office. The young Marines always worked outside in the Okinawa heat. Something was wrong, I thought. Why would he want me in his office?

"Corporal LT, I want to talk to you about your girlfriend, Mona," he said with a sincerity I will never forget. "You have yourself a jewel, Marine."

"I have been in the Corps for 22 years and have met many girls around the world, but my wife and I know that she is something special," he said with a soft voice and a quiver in his lip. "Don't let this one get away, Marine."

Although he said these things, my thoughts at the time were not marriage or long-term relationships. Even more so, I thought that marrying someone from another country was out of the range of "happening."

A few months later, I headed back to Albany, GA on leave determined to get out of the Corps and go to college. While there, I hung out with my high school friends and wanted the good old days back. I thought it would be a great time and all those who didn't really know me from high school would think I was "cool" now and would let me become part of their world.

While in the car with one of my old friends who was not a

believer, I had a thought that would forever change the course of history or me. "I miss Mona."

I couldn't get her out of my mind. My plan to get out of the Corps was solid, yet for some reason living without my friend, Mona, brought a sense of emptiness that couldn't be explained.

I was specifically thinking about how she helped me and other Marines understand a Japanese weekly drama called "Tokyo Love Story" where a simple working man proposed to a beautiful woman each week. In Japan it is known as "Hyaku Ikai Tokyo Proposoru" or "The 101 Tokyo Wedding Proposals."

Before we officially started dating, I told her that I was explaining to the Marines in the barracks about this drama that really captivated me. When she came in to help me translate, she was laughing so hard because my translation of the story was completely wrong. The Marines were like "Hey, Cpl. LT, we thought you knew what they were saying?"

I had no idea, but I just loved the Japan dramas.

From this particular show, there was a famous actor named Eguchi Yosuke (A-gooch-e Yo-soo-kay) who had very long hair. I liked his personality and coveted his hair, so my future wife started calling me Yosuke, which has stuck to this day.

So, I have a Japanese name, Yosuke, and she has an American name, Mona.

She got her name from her professor in college who called her Mona because her best friend was Lisa. He couldn't remember her Japanese name, so he said from now on "you are Mona because you sit next to Lisa."

Reflecting on how much fun we had together, I knew she had to stay with me.

My Proposal

When I flew back to Okinawa, she was there with a smile, some flowers and wearing a nice outfit. A few weeks later I took her to the place we first met and got on my knees and asked her if she "would spend the rest of her life with me."

I approached the Staff sergeant detailer the next day and told him that I wanted to reenlist in the Marines because I was getting married. He said, "I wish you would have come to me a few months ago because the 7242 Air Support Operations Operator field is closed for the year."

"There are only two options now," he said. "Brig guard and this thing called Public Affairs."

"What is public affairs?" I asked.

"It says here that you have to write stories and take photos for newspapers," he said, as he picked up the phone and called the Marine Corps Base Camp Butler public affairs office (PA).

"Can I bring a corporal over to interview for a PA position," he asked in a professional voice. "Tomorrow at 1400. Ok, I will bring him over."

I called Mona to tell her what was going on and I asked her if she knew anyone who could sew name tapes on my uniform because that week, the Marine Corps put out a Marine Corps order stating that all Marines need to have name tapes on their Battle Dress Uniform and had one year to get it done.

No one had name tapes on their uniform at that time. Again, the order had just come out that week. I rushed the uniform over to my future wife, and amazingly she had my uniform starched and name tapes placed on it just in time for the PAO interview.

Walking into the PAO office, the giant Master Gunnery Sergeant Public Affairs Chief said "Corporal, where did you get those name tapes?"

"I have connections Master Guns," I said with a smile shaking his hand.

"I like you already," he said as he proceeded to hand me an English Diagnostic test.

A few hours later he welcomed me to Public Affairs and already had orders for me to go to Parris Island, SC where I would work at the Boot Newspaper as a journalist.

If I had not met Mona, I would not be writing this book and there would be no *And We Know*. Everyone tells me they love the prayers at the end of each of my videos. She is the one who urged me to pray at the end. I told her "no" many times thinking that no one liked them. SHE WAS RIGHT AGAIN!!! She is the backbone of *And We Know*.

God was Building "A Good Foundation."
This year (2023) my bride and I entered our 31st year of marriage. We never envisioned our future on that morning of June 2, 1992, as we stood inside our 1965 mobile home. We had no idea that we would be blessed with seven children, though we will always carry the pain of the loss of two of them that died in the womb.

I recall our marriage day like it was yesterday. A man with a long ponytail and old pickup truck asked us to fill out some paperwork, and after we paid him $70 for the marriage certificate, and he stood up to walk out.

"Hold on," I asked the unshaved man who had a deep Southern drawl. "Is that it? Aren't we supposed to say, 'I do?'"

"Sure," he said with a chuckle. "I have just the thing," pulling out a wrinkled piece of paper with some spilled coffee on it.

As he proceeded to have us repeat the vows, we laughed and found that we couldn't finish them. It all seemed like a dream. We were shy, young, and very naive. Although it was a strange beginning, my mind and heart were determined to make it 'til death' do us part.

Eighteen months later, Mona and I flew to Okinawa to have a proper wedding ceremony in front of her entire family. Seeing her in that bridal gown was one of the best things we ever did for our marriage. In order to pull this off, we both had to work extra jobs to save up the money for the flights and the wedding. During the middle of the reception dinner, every Okinawan man who attended stood up, walked up to our table and placed an envelope in our basket and bowed. Mona never told me about this custom. She laughed as we opened the envelopes and discovered that we got back all of the money we invested into the wedding. God was taking care of us, and we didn't even start to realize it til later in our marriage. Well Glory.

Chapter Three

The Good Drive!

If ye then, being evil, know how to give good gifts unto your children, how much more shall your Father which is in heaven give good things to them that ask him?
[Matthew 7:11]

I would always tell Christian Marines through one-on-one mentoring that we should be excited about receiving PCS orders. No matter where we are assigned, we know that it will be a ***good*** place, as God doesn't make mistakes.

This happened many times in my journey through the Corps, but two years into our marriage, we got our first dose of God's orders.

"You have orders to New Jersey," said the Master Sergeant over the phone.

"I asked for New Mexico, Master Sergeant," I said in disbelief while sitting in my desk at Parris Island. "That is the last place on earth we want to go."

He didn't flinch. My cries of desperation for new orders fell on deaf ears. Most Marines…I take that back…most veterans know exactly what this means. Take the orders. You have no choice.

We were assigned to support recruiters through a position called Public Affairs and Marketing. We essentially were responsible for putting events together with companies in the area to bring "leads" for the Marine recruiters. We also had to visit radio and TV stations to ask them to play our Public Service Announcements. By law, at the time, they were required to play a certain number of PSAs each year.

It was difficult breaking the news to my wife who was pregnant at the time. All the books through life told me that the Northeast of the United States was a scary place. There would be gangs, people would steal our car, my wallet had to be hidden and more. Please don't be offended if you live there.

That's just how you think when the media and books infect your mind.

An Eye-Opening Northeast Migration

We made our trek from Beaufort, SC to Naval Weapons Station Earle in a small 2-door car. When we finally made it to New Jersey (NJ), we pulled into a rest area on the NJ Turnpike in pouring rain. We will never forget the sharp answers from the "natives" to our questions.

"What do you want?" asked the elderly lady behind the counter in a very harsh tone.

"We just wanted to know what exit we should take on our way to the Navy base," I asked in shock to the tone of her question.

We continued our drive, while in shock at what we thought were very direct and cold responses to our questions. We arrived at a small hotel next to the base. The man behind the counter was wearing a turban and acted as if he was from another country. We walked into the room, sat on the bed that was leaning slightly due to a broken leg, looked at our son and we wept.

I apologized for bringing her to this dark and dreary state. Remember, we were brainwashed into believing what the media, books, and others told us. Add the fact that growing up secluded in the Deep South away from civilization can really add to those misconceptions and feelings of fear.

The next day we woke up to a bright sun and clear skies. Hunger kicked in so we drove around looking for a restaurant and ended up seeing a building that resembled a model you would see on the set of *Happy Days*. It was called a diner.

We walked in and had a seat. To our surprise, the waitress came over to greet us and did something you would never see down South.

"Oh, look at the cute baby," she yelled out as she reached down and grabbed our son. "Look everyone. This baby is so adorable," as she proceeded to walk around the diner.

She didn't ask permission to pick him up. We were just a young couple in a strange, new world who cried most of the night wondering where it all went wrong. You can probably imagine our shock at such an experience.

A few minutes later, everyone in the restaurant was walking over to meet us and find out where we were from. That broke us out of our comfort zone and opened our social "shell" we loved to crawl into.

We fell in love with the Garden State. It was wonderful. There were three shopping malls nearby for my wife to enjoy shopping, New York City was just down the road, and bagel shops everywhere. But the greatest part was the mix of so many different people and cultures packed into one area.

It was here that I learned the value of loving my wife as Christ loved the church. He died for the church. That impressed upon me my need to die for her in such a way that my life and wants were not as important as hers. That I was willing to sacrifice all for her, just as Christ sacrificed all for His Church (Eph. 5:25).

The difficulties in marriage were starting to increase again, and I knew that attending a local Bible study would help me. The Holy Spirit has an amazing way of drawing you to the Word of God when things aren't going right. I fell in love with BBN radio and couldn't get enough of the sermons each day during lunch and at night from 6:00 to 9:00 p.m.

My wife was not a believer yet, but she knew there must be a God after several "unbelievable" things happened once I started reading the Bible and telling her about it.

Our "Miracle" New Car Purchase

One such incident was when we needed a 4-door car because our family was growing. The 2-door was becoming harder for us to navigate with car seats and more. New Jersey had a plethora of used-car dealerships to choose from. They had a magazine filled with a variety of great auto sales in the state.

My wife has always had the mindset that we should look at all options available before buying anything. When we met, I would walk into a store, pick a shirt, buy it, and walk out. She taught me the value of "Clearance" sales.

Well, we started early, driving from one car lot to the next throughout the middle of the state of New Jersey. We were excited walking into each dealership, but we left in pain and disappointment each time.

You see, we paid full price for our new 2-door car at around $12,000. The most we could get for a trade in was about $6,000. As a sergeant in the Marine Corps, those numbers were not looking good.

Add to this story the fact that I wanted to get home in time to watch a huge golf tournament on TV. There was no way I was going to miss Tiger Woods, I thought.

All day long, our pattern was the same. Open the magazine, find the car, drive to the dealership, watch the guys checking out our little car and look at us with pity, and walk in to give us the bad news "We can only give you $6,000 for your trade in."

By mid-afternoon we pulled over to eat, exhausted from this car-shopping nightmare. As I picked through my food, I was thinking about the golf tournament when the unexpected happened.

"Look, we must have missed this one," my wife said with that look. You know the one. The sad eyes moved toward me with a magnet tugging on my heart.

My mind was whirling. Tiger vs. my wife. My heart was beating so hard, I could hear it. My blood pressure was going up because deep inside I wanted to say "No, we must go home. Let's give up, plus golf is on right now."

But the Bible verse I read that morning was nudging at my soul, "Husbands love your wives, even as Christ loved the church and gave Himself for it" (Ephesians 5:25).

My mouth opened to say no, but the words that came out astonished me. "That looks great. I will call them to see if they are open and the car still available."

I walked out to the phone booth. Remember those?

"Sir, hello," I said with a sad voice. "We were looking at this Mercury Sable you have listed at $12,000…do you still have it?

"We sure do," he said. "Come on over. We are open for another two hours."

My wife and I jumped in the car with the two boys and made the 30-minute drive to the dealership and walked in.

The car salesman saw that I was a Marine and told us that the dealership owner was a Marine also. It was all great until they took the keys from our car and proceeded to look over our brand new 2-door Toyota (on which we still owed $12,000).

"Uh oh, here we go," I said to my wife. "As they walked in to give us the "bad news" we actually stood up and prepared ourselves to apologize for wasting their time. We knew what they were going to say—so we thought.

"Where are you going? Have a seat folks," the dealer motioned with a smile.

"We know it is a bad trade in sir, and…"

"Hey, no need to leave," he said. "We talked to the owner and worked out what we think will be a great deal for you."

As he looked at my toddler sons and looked into our eyes, he made a quick glance to the back office.

"We would like to give you $9,000 for your trade in (which reduced our loan to $3,000) and reduce the cost of our car to $8,000. That will put the total cost for the car at $11,000. How does that sound," he said with a wink. We both cried with joy!

That moment in time will never leave us.

The best part of the entire ordeal was that I finally did what my wife asked without complaining. The reward from that was more than saving money, it was as if that dealer was an angel playing the role of a messenger to provide the good news. It was the break we so desperately needed.

Add to that the fact that the owner knew we had a long drive home and that a storm was coming so he directed us to follow him. He knew a shortcut.

We were filled with joy. That was a *good* drive.

Chapter Four

A Good Amen!

Then I heard something like the voice of a great multitude and like the sound of many waters and like the sound of mighty peals of thunder, saying, "Hallelujah! For the Lord our God, the Almighty, reigns."

[Revelation 19:6]

Many great things occurred in New Jersey that made me understand how wonderful the Word of God really is, especially when you read it, pray for increased faith and watch "unbelievable" things occur.

In my opinion, nothing is greater than watching someone pray to receive Christ.

Remember, my wife and I were in the early stages of going to church, reading the Bible and learning to pray together for the first time in our marriage.

The first thing that we witnessed was what most would think is small, but for my wife, who was still questioning if my faith in Christ was real, it was huge.

The word tithing came up. Giving ten percent of our income to the local church seemed strange. Our income was low, we had car payments, raising kids, and so much more.

I ran to the bank on payday, withdrew ten percent of our income and rushed to the chaplain.

"What is this," he asked with a smile.

"Sir, I must give this to you now because I know we will spend it this weekend. We don't want to miss out on the blessings from this," I said with a feeling of joy that I had not experienced for many years.

> *'Bring the whole tithe into the storehouse, that there may be food in my house. Test me in this,'* says the Lord Almighty, *'and see if I will not throw open the floodgates of heaven and pour out so much blessing that there will not be room enough to store it.'*—Malachi 3:10

About two weeks after we gave our first tithe, in October, the IRS sent us a letter. To our amazement, there was a check for around $1,300 for Earned Income Credit they owed us.

"You have got to be kidding me," I said in disbelief. "Could this be real. It has to be because of the tithe." My wife pondered this in her heart.

So, you can imagine how much I praised the Lord.

Everywhere I went, listening to preaching on the radio, reading the Bible at night with the boys, feeling a love for Jesus Christ that I had never before felt.

While working in the recruiting station, the main office Marines were responsible for driving possible future Marines to the Military Entrance Processing Station (MEPS) in Philadelphia.

It was a long 90-minute drive once a week from Iselin, NJ to the heart of Philly.

"Excuse me sir, is that a Christian radio station you are listening to," asked the 18-year-old recruit in the passenger seat of the military van.

"Yes," I said thinking he wasn't paying attention.

The young man in the back seat sat up and laughed when the passenger said "You are a Marine and a Christian? I thought you weren't allowed to be a Christian in the Marines," he said with a confused look on his face.

Now, there are two thoughts here. Do I have a conversation about this, or do I support his view that we can't have this conversation because he might be a Marine someday and this is government time? There was no law stating that I was not allowed to share Christ with anyone, yet there was a thought that proselytizing during working hours could be a problem.

Before I continue with this story, it is necessary to understand my thoughts on military folks serving Christ. This has been a hot debate for many throughout my career and there is a simple biblical answer for this in Matthew 8. Incidently, this is the same Scripture often misused by those in pseudo-religious groups known for going door-to-door sharing their unbiblical message.

In this biblical story, a centurion approached Jesus desperately pleading for Jesus to heal his servant who was afflicted with palsy. Jesus said he would "come and heal him."

Instead, the military leader acknowledged that Jesus has great authority (like himself, but much greater) and he had so much respect for Jesus and belief in His ability to heal his servant without even coming to his house. He therefore boldly requested that Jesus simply give the command and his servant would be healed.

Now, when speaking with folks who don't believe that Christians can serve in the military I get to this part of the story and say something like this (with a touch of sarcasm, of course), "You know what Jesus said when the centurion told him this? Jesus said, 'Before I heal your servant you must first take off your uniform, become a civilian and then you can follow me.'"

To which a typical reply would often be, "Really?" Followed by, "Oh, wait a minute, that's NOT what Jesus said. Wow."

I would then pull out my Bible app, go to the Scripture passage and read the following words from Jesus:

When Jesus heard it, he marveled, and said to them that followed, Verily I say unto you, I have not found so great faith, no, not in Israel.—Matthew 8:10

I always cherished these opportunities to share with folks how the centurion in this Bible story (the military leader), had many men under his authority, and they did what he ordered, but he was humbled by the power and love he saw and recognized that HE was under Jesus' authority—so much so, that he did not think himself worthy that Jesus should even set foot in his home.

During one such group encounter, I asked, "Can I pray for you?" to which one girl responded "No, that's ok. We pray to different gods."

The Christian Radio Station

So, while we were in the van one of the young men was astonished that I was wearing my Corps uniform and listening to a Christian radio station.

"I love Jesus Christ," I exclaimed while driving. "He is the reason for living. Without Him we are nothing on this earth."

"Well sir, I thought that we are living in hell right now and that someday we will pass on to a better life," said the young passenger.

The guy in the back joined in. "No man, we simply die and come back as someone else. I am in my 8th life right now," chuckling as if he knew it was all a joke. Of course, we know this is unbiblical as he believed in reincarnation.

If they had questions, I felt obligated to answer them. No one said we weren't allowed to ask questions. The Marine Corps didn't put out a manual stipulating that chaplains were the only ones allowed to share their faith. As a matter of fact, General Krulak, the Marine Corps commandant at the time, would share his faith at dinner parties. I saw them on video at the time. More on that in a later chapter.

The "One-Week Challenge"

So, the strangest request sprang out of my mind, through my vocal cords and directed at the two young men in that van while on the NJ Turnpike.

"You should give Jesus Christ a try for one week. If you don't like what you get, then tell him to leave you alone and try something else. You won't regret it," I said with a smile.

Now that moment is frozen in time for me. After sharing the love of Jesus with them for 45 minutes, talking about different Bible verses regarding military and Christianity and so much more…thus, the "One-week Challenge" was birthed. As we were entering Philadelphia, I asked the young men to read the small military Bible I gave them and talk to Jesus for one week and if He wasn't who He claimed He was, they would know.

We pulled up to MEPS Philadelphia, I said a short prayer and they walked into the building.

Crazy, right? Here I am, a 26-year-old sergeant, reading the Bible, singing His praises all the time, watching my family slowly change and the next thing you know, I'm talking about Jesus to a couple of young folks in a van.

After a long drive back to the recruiting station, and then another long drive back down the Garden State Parkway to exit 120, my little truck finally made its way back home and slowly rolled up the house.

After getting settled, exhaustion set in. The normal routine of listening to the sermons on the radio put me to sleep.

"...please forgive me of my sins," I woke up to a man's voice speaking those words into my ear as if it was coming from an old record player, "in Jesus' name. Amen!" I heard him clearly as if he was right next to me.

Right after he said the word "amen," a large voice filled my entire room with a shout "AMEN!" It was so loud, yet it didn't hurt my ears. Then from the lower part of my bed, a sort of scanning process of my body occurred as if I was literally on a printer/scanner. My body felt the power surge of what I now know was the Holy Spirit. It started at my feet and worked its way up to the top of my head.

I felt cleansed and as if I was in heaven with a Holy Spirit sensation that can't be put into proper words.

Slowly, my head turned to look at the clock next to the bed that showed 2:00 a.m.

"Was that you, Lord?" I asked. "Hey," bumping my wife. "Did you hear that?" She was sound asleep.

I ran downstairs and prayed and then ate a small meal, turned on the TV to get my mind off what just happened, but nothing helped.

Suddenly, it all made sense. "The two young men in the van," I exclaimed aloud to myself, knowing it had to be the voice of one of those two guys.

So, off to work I drove. However, on this day we were required to attend a recruiter gathering at one of the hotels in New Jersey. These were very important monthly conferences that allowed the recruiters to get a breath of fresh air away from the daily grind of finding another "body" to fill their monthly quota.

The business of taking photos, writing a news story, talking to recruiters and more kept me from doing the one thing that needed to be done, ASAP—call those young guys to find out what happened…hopefully to find an answer to my "amen" experience the night before.

Finally, I had an opportunity to get to a pay phone in the lobby of the hotel.

"Hey gunny, I was wondering if you have the number to MEPS Philadelphia," I asked knowing he wouldn't give it to me. He was the last guy at the RS (Recruiting Station) headquarters I wanted to talk to.

As I grimaced and waited for his not-so-delightful answer, the Gunnery Sergeant asked me "Why do you need it?"

"Well, there were two young guys in the van, and I just needed to ask them something."

"They're right here," he said and then he handed them the phone.

"Hey, do you remember me," I asked the guy who was the one sitting in the back seat who believed in reincarnation. "I was wondering if you followed through with my suggestion and prayed to God?"

"No sir, you said to wait about a week and give it a shot," he said without pausing.

"Ok, can you give the phone to the other guy," I said with disappointment.

"Hello, I was the guy driving with you guys yesterday down to Philadelphia and we had a great talk about faith. Do you remember me?" I asked thinking he wouldn't remember.

"Yeeeessssss sir," he said with the longest "yes" in history.

"Did you happen to say a prayer last night or early this morning?" I asked.

"Yeeeesssss sir," he slowly answered again with an inquisitive tone.

"You did!" I yelled into the phone with excitement. "Can you tell me what time that happened?"

"Well sir, it was really late…it had to be around two o'clock in the morning," he said with a puzzled sound to his voice.

"You did it! You did it! Way to go!" I yelled into the phone, totally forgetting that I was standing smack dab in the middle of the hotel lobby with a bunch of Marines and civilians. "God has a great plan for your life. I know it. He allowed me to hear your prayer young man. He wanted you to know that. You are saved by the GRACE of GOD!"

I hung up and didn't care that everyone heard me.

After checking into the hotel and getting the key, I walked into my room and noticed a Bible sitting on the bed. We each had to share a room with another Marine at the conference to save money.

I turned around and sure enough there was my friend from Parris Island. His life turned around also as he recently came to know Christ as his Savior.

We talked all night long about all that had transpired in life and celebrated the recent story about the young man who prayed the night before.

It was certainly a really GOOD "AMEN."

And Jesus said unto the centurion, 'Go thy way; and as thou hast believed, so be it done unto thee. And his servant was healed in the selfsame hour'.—Matthew 8:13

CHAPTER FIVE

Good Faith to Move Huge Mountains!

Truly I tell you, if you have faith as small as a mustard seed, you can say to this mountain, 'Move from here to there,' and it will move. Nothing will be impossible for you.

[Matthew 17:20-21]

The title I had while working in marketing for the Corps at Recruiting Station New Jersey was Public Affairs Non-Commissioned Officer (PANCO)—we called it "PANCO Duty." We had a basic manual describing our mission, but the Marine who I replaced created a database program to help the station track data on recruiting in that area.

In a nutshell, as long as we were getting results to help the Marine recruiters find "bodies" for the Corps, we were doing a good job.

We had VCR tapes with those famous 30 second and 60 second Marine Corps ads, created by the J. Walter Thompson agency, that every Marine loves. We also had amazing representatives from that company who would share better ways of marketing and getting "leads" to help recruiters find future Marines.

We often drove into Philadelphia with our contact cards for all of the radio and TV stations, walk in for a "cold call" and see if they had a few minutes to discuss the Public Service Announcement (PSA) and why it was important for them to support the U.S. Marine Corps. One of the perks of the job was getting free Philly bagels and coffee.

We called ahead for one of those meetings. That was a mistake. The cable station heard the "U.S. Marine Corps" was coming, and therefore assumed we were bringing them a possible contract for millions of dollars. So, they brought along the best sales reps from the state.

When I arrived, they had four decked-out sales reps and the head of the station waiting at the door to greet me. That was the one time I felt like Donald J. Trump.

"How are you doing today, sir?" they asked.

"Please don't call me sir, I work for a living," I answered with the standard enlisted Marine answer. Their laughter filled the entire building as they were pouring out love on a scale I had never experienced before.

As we sat in the huge meeting room, they dimmed the lights to show me a special video presentation about their station and how they could reach our target market of 17 to 20-year old prospects throughout the area.

"Now we get to the good stuff, sir," the sales rep nervously said as they directed me to open the U.S. Marine presentation folder. "Let's talk about the different price packages."

"For $650,000, we can reach your target market in every household with your ad for the next six weeks," the sales rep said with ease.

Big gulp. Sip my coffee. Blood rushing and pounding through my body as I think, "how am I going to explain that I am here to ask for this ad run for FREE?"

After a few minutes, I graciously interrupted and told them the truth about our PSA request. Phew.

Each meeting was nerve-wracking to say the least as it was way out of my comfort zone, but every now and then I would get a huge break through. On another occasion, a billboard company director that really loved the Marine Corps said he was happy I drove in and promised to have 150 billboards up throughout New Jersey for us. There was nothing better than letting your command know about these sort of wins.

Our Heavenly Father is Amazing!
We had so many side jobs which included driving around fu-

ture Marines, working the Marine Corps Ball annual event in NJ, creating certificates for award ceremonies, developing and managing databases, writing news articles and creating a newsletter once a month, and so much more.

New Jersey had so many opportunities. We worked every year with the "Garden State Games" to bus in the top athletes from around the state to compete in several track and field events that were similar to what the Marine Corps does in training—ups, sprints, sit-ups and more.

We would set up meetings for the next-year's event one week after the event was complete. It was that big.

We had to arrange dates with the schools, ensure the Marine Corps could provide the recruiters to help monitor each event, work with the Marine Corps League for meals for each athlete, coordinate with an I&I station (Inspector & Instructor) to have military vehicles park nearby and more. The best part was orchestrating a CH-53 fly in for all the students to see.

Then we heard the bad news. "We are looking at what looks like a Nor'easter for the weekend. It's not looking good for the Garden State," said the meteorologist on a local television station.

This was the forecast leading into our huge event where we expected 1,500 students—the largest event we have ever organized.

"This can't be. We have been working on the Garden State Games event for so long. Why Lord?" I pleaded in prayer.

Then, the strangest thing happened after talking to my Heavenly Father. PEACE.

The storm picked up strength through the night. The rain was pouring out like that of a monsoon in Asia. I kept thinking about the disciples' lack of faith as they went to wake Jesus when he was sleeping in the belly of the boat and ask for help at a time when they thought they were going to die during what must have been an enormous storm.

> *The disciples went and woke him, saying, 'Master, Master, we're going to drown!'*
>
> *He got up and rebuked the wind and the raging waters; the storm subsided, and all was calm. 'Where is your faith?' he*

asked his disciples. In fear and amazement, they asked one another, 'Who is this? He commands even the winds and the water, and they obey him.'
—Luke 8:24-25

I didn't want that sort of rebuke. No way. So, to get through and believe that the storm would go away, I drove to the other side of New Jersey, from Naval Weapons Station Earle to Fort Dix and McGuire Air Force Base for the event with a smile.

I was singing hymns from the top of my lungs while the rain was crashing down on the windshield. The music was drowned out by the roar of the storm. The windshield wipers couldn't move fast enough. It was a dangerous drive, but my Heavenly Father gave me peace.

As I pulled up to the event location, there were a few buses and the RS New Jersey sergeant major got out of his vehicle to talk to me. His giant 16-inch arms on that 5-foot 4-inch frame were intimidating as they reached down to my car window.

"Not sure if this is happening today sergeant," he said with a seed of doubt in his voice.

"Sergeant major, it's going to be a good day. Don't worry," I said with joy and confidence.

He looked at me puzzled and said, "oh, you are talking about God," with a chuckle.

"Yes, sergeant major."

About 30 minutes later, the rain stopped. All of the buses started pulling in. Marines were working feverishly to get the course set up for the day's events and the Marine Corps League retired Marines were setting up their tent. We were quite busy. So busy, in fact, that I had just a moment to look up to see all that God was doing, but that's all it took.

A strong ray of sunshine blinded my eyes as I looked up into the sky.

"Do you see that, sir?" the question leaped from my voice as if I had just won the lottery. "Look at the bowl of blue sky! Isn't that amazing!"

The retired Marine just stared at me for an extra-long moment like I was crazy and kept working.

The sky looked like God had taken a clear bowl and placed it right over the base. There was a circle of dark clouds that appeared as if they wanted to move into the area directly above us, but there was no getting in.

Then we heard what sounded like thousands of horses running through a field in the distance. All of the sudden, a Marine Corps CH-53 helicopter appeared in the sky, making its way onto the field about 300 yards away.

I sprinted to the CH-53, jumped on board and asked the crew chief how they were able to make it because I assumed their flight would be canceled due to the storm.

"Sergeant, it was the strangest thing today," he shouted through the sound of the helicopter. "There was a huge opening in the sky over our area in Pennsylvania, and we saw the opening for your area and knew we could make it."

"That was God," I said with a shout.

The event went as planned. Every single athlete participated with no rain whatsoever. As the final student sprinted to the finish line and the Marine yelled out his finishing time, a single rain drop hit my head.

Everyone rushed to get their tents packed up, and then buses started pulling out. There was no time for chit chat. As the final bus pulled away, it started raining hard. What an amazing day.

God's "Signature"
Not so fast. Our Heavenly Father is so amazing. He does what I have called for years now—"His SIGNATURE." It seems that when we are in the Word of God, He speaks to us and reminds you and me that, as we put our trust in Him, He cares and orders special circumstances that touch our lives in memorable ways.

Each night, I made it a habit of reading my Bible while my two toddler sons, Brandon and Anthony, watched and fell asleep. These were such special times. The boys enjoyed my prayers (early scenes of what has come to be a highlight of my *And We Know* broadcasts), and there is nothing like your sons falling asleep with their dad right there next to them. You know, being a dad is a highlight of my life and one of the most important God-given roles I deeply cherish.

One night, I simply opened my Bible. It fell open to the Matthew, chapter 17.

> *Truly I tell you, if you have faith as small as a mustard seed, you can say to this mountain, 'Move from here to there,' and it will move. Nothing will be impossible for you.*—Matthew 17:20-21

Mustard-Seed Faith

According to several scholars, the mustard seed is about 1 to 3mm in size and one the smallest of all seeds when Jesus was speaking to those who understood this. The amazing thing about this seed is that it can start growing the day after it has been planted.

It is a small seed that grows very fast into a large plant, without the need of cultivation, and can be seen in the desert—some even near the Jordan River.

So, we see a small amount of faith in something that seems impossible like Jesus' metaphor of "moving mountains" and when that "mountain" or impossible situation is moved, we see the evidence of the large fruit that comes from the fulfillment of that seemingly impossible situation.

We experienced such a "mountain," during those Garden State Games, when we watched what appeared to be a Nor'easter move around a field to allow the games to go on, and a few thousand people to see the handiwork of God. This so often results in encouragement for the one in prayer to see a divinely appointed miracle and prepare the hearts of doubters to see His Majesty in more ways when encountering similar experiences in the future.

This would be the Sergeant Major. The next "mountain experience" is about to unfold. Stand by!

CHAPTER SIX

The Taj Mahal was Good!

Verily, verily, I say unto you, He that believeth on me, the works that I do shall he do also; and greater works than these shall he do; because I go unto my Father.

[John 14:12]

This job in New Jersey was certainly becoming one of adventure that would challenge every fiber of my Marine Corps life, but more importantly, my spiritual life.

God has a way of directing our paths that are sometimes very uncomfortable.

Trust in the Lord with all thine heart; and lean not unto thine own understanding. In all thy ways acknowledge him, and he shall direct thy paths.—Proverbs 3:5-7

Why did the recruiter sign me up for Aerial Navigation? Why am in Public Affairs now? Why did He make sure my wife appeared on the scene at just the right moment? Why send us to Parris Island and then to New Jersey? Why am I constantly working and learning a new job in the Corps nearly every single year? Clearly, I had a lot of questions, but I was also learning a great deal about God and his loving ways—even though, at times, they didn't feel so loving.

Marine Corps Ball
"As a PANCO, you are responsible for organizing the Marine Corps Ball for 250 people every year at the Radisson Hotel," said the PAO staff sergeant who was on his way to Okinawa. "No wor-

ries. It's not that hard. I created a turnover file to ensure that it runs smoothly for ya."

That's it. He provided a quick rundown on how this entire job works for New Jersey—*in one day.*

"It's all yours," he said, hurriedly finishing our turnover so he could get back to checking out and packing his furniture for the next move.

Sitting in a small office, I stared at a computer with this new system called "e-mail," which was slowly filling my "inbox."

The phone was constantly ringing, recruiters were trying to become my friend so that they would get free t-shirts and other giveaways for their upcoming events—the admin department needed driver duty info, the executive officer kept pushing us for meetings, and the station commanding officer kept threatening to send us out on weekends to help recruiters if the quota wasn't met.

That workload piled up each day, but nothing compared to the Marine Corps Ball. Most balls are done by an entire group of leaders throughout the year. I had to put the entire event together, ALONE.

The Marine Corps Ball is one of the most important events in the life of a Marine. It began in 1921 when the commandant Gen. John A. Lejeune issued an order that the birthday of the Corps would be celebrated each year with a ceremony.

In my experience over the years, many Marines who are first-timers to the event, complain about it, believe it or not. They don't want to pay for it, or have expressed that they have better things to do with their time. But once they've had a chance to attend the Ball, everything changes, and all the complaining disappears. Every Marine is made to feel special and filled with American pride, remembering why they serve their country.

Every Marine Corps Ball has some special traditions that holds some great memories for us Marines. At one point during the event a video is played featuring the commandant and sergeant major of the Marine Corps. In addition, the cake with the birthdate of the Corps, November 10, 1775, along with how many years it has been active, is marched into the ballroom in slow motion as the Marines' Hymn plays and we join in together in the celebration.

There is also a standard speech given by the emcee and the oldest Marine eats a piece of cake and hands a piece to the youngest Marine present in the room. This signifies the rich Marine Corps tradition that is being passed on through each generation.

Event venue, ticket sales, event theme, ticket design, dinner selections, cake, DJ, newsletters, guest speaker, hotel room reservations, transportation, color guard, and MWR support were just a few of my responsibilities.

The first event I was responsible for putting together turned out great. We were able to use the same hotel in the middle of New Jersey as the prior year. The hotel management and staff were very professional and helped me so much with the event logistics and special details expected by the Marine Corps.

That first event wasn't without some pain, though.

Everything was running just as planned. The recruiters were pleased, and their wives were dressed formal as if they were attending a wedding; with perfect hair, nails done, bling everywhere and so much more. But most of them, in my opinion, showed up because it was their "duty" as a spouse.

You see, one of the things that I didn't understand, having been with the station for only about six months, was that these recruiters would work from early morning to nearly midnight, seven days a week. As a result, many marriages were falling apart.

The stress on the recruiter was 100 on a scale of 1 to 10. If the recruiting station did not bring in the numbers needed for that particular month, every Marine was on edge.

One of the things that delighted me a great deal was the friends I made who love Jesus and made it a priority to make recruiting a teamwork event. Each spouse would bring food to their husband with their children in tow. It paid off in many ways. The young folks had the opportunity to see that most Marines have families, subsidized housing, and everyone seemed to be in a good mood around Marine dad.

The other payoff—the stress typically experienced by Marines was reduced due to the support of their family. They didn't have to hear "When are you coming home?" on the phone again while trying to teach and be an example for students on how they could serve their country.

[Back to the Marine Ball]

"Sergeant, can I talk to you in the hallway," said the hotel event coordinator with panic clearly expressed on her face.

"Sure," I said, glancing at my wife the other seven folks at our table with confusion.

As I walked into the hallway with the event coordinator, she said, "we have 150 chicken orders and 50 beef."

"How can that be? Each table has a nametape with their order on it," I responded, realizing I did my part in order to prevent this from happening.

"The waiters asked each person what they ordered because every time they put a plate in front of the customer, the customer would say, 'That's not what I ordered,'" she said with increased anxiety.

I told her to please place the food on their table aligned with what was on their name card and not to worry about it.

As I walked over to sit next to my wife, several wives stood up from their table. It was like a slow-motion movie, watching each Marine look at their wives as they would slowly point in my direction. One wife after another would snap her neck toward me with a piercing glare, then stand up and walk towards me. You get the picture?

Within 30 seconds, eight wives were standing in line to talk to me about their food order. This was an unusual response, but again, after spending three years in recruiting you discover why they were looking for anyone to blame for their difficult lives.

Suddenly, the microphone made a screeching sound throughout the ball room—the kind you hear in movies when the band leader tips the microphone over and everyone covers their ears.

"I have been involved in dozens of Marine Corps balls and have never seen this type of behavior!" screamed the 6-foot, 3-inch, skinny white sergeant major. "You will eat what they put in front of you and you will like it!" he ended as he slammed the microphone on the podium and walked through the room to the exit.

The following year, I made every single Marine sign a document on behalf of their spouse and fax it in. Only one wife walked up to me to complain about her order that year. I pulled out my file, showed what her husband, a master gunnery sergeant, had signed. She said, "Oh no, he didn't just send me over here for that," as she did a 180 and apologized for bothering me.

The Impossible Can Happen!
When I was handed the mantle to run the Ball at the recruiting station, aspirations of having the biggest and best Marine Corps Ball in the USA started buzzing through my mind. My wife and children will tell you that when it comes to "vision" I will often choose what many perceive to be impossible and   talk about it for days.

Well, I had to visit the recruiting station in Atlantic City a few weeks after arriving as the new PANCO. Of course, anyone who drives through such a well-known city wants to see the Strip. The huge building called "Taj Mahal" on the Strip caught my attention, so of course I pulled in to have a look.

As I navigated through the massive halls and corridors, I was somehow able to make my way to a section that provided prices on holding events at the Taj Mahal. My initial reaction of excitement was dashed, though, when I saw how expensive is was. Our ticket price would have to increase significantly to about $100 per person, which was quite a lot for a young sergeant with a family.

We were charging about $50 per person at the time which covered meals, etc. MWR money would cover $15 per person to ensure it was not a burden for couples.

There might have been a prayer slipped in when driving back to the office that sounded like this "Heavenly Father, I would love to have the Ball here. You could make it happen, but I submit it all to Your will."

Two Years After My Taj Mahal Prayer

Well, two years later, wouldn't you know it, we were planning the Ball at an exclusive venue in the Atlantic City area (not the Taj Mahal. But hang in there, you are about to see somewhat of a "miracle").

"Sir, I had a meeting with a hotel in Atlantic City," I said to the Recruiting Station commanding officer who was a Major (I will leave the name out of this book to protect the identities of the folks involved in this). "They will have a large building ready for us next year."

Of course, by this time, the CO, XO and sergeant major had a deep trust in my word and allowed us to move to a new hotel. It was a bold move seeing that our event was attractive and comfortable at an amazing venue up north.

This hotel we had chosen hosted many events throughout the year for each branch of the military. They hosted recruiters, the Marine Corps Scholarship Golf Tournament that included the Silent Drill Team, high-profile contractors from Washington D.C., and so much more.

The manager had a vision to have a huge venue built behind the hotel for special events, so he called me to go over our event and we had a wonderful meeting. Their team presented an impressive sales pitch with a large print of the future ball room for us to see.

Each hotel room would be reduced to $25, and there would be a childcare facility set up in a recreation room downstairs for the children of our guests. They were also including free wine for all and flowers for each spouse.

How could this go wrong? In my younger years, I had a habit of trusting everyone, and in this case I had no reason not to as everyone seemed professional and sincere in fulfilling all our requests for the event.

Eleven months later, we were ready for the Ball.

While in a meeting with the staff and the Sergeant Major, who was the highest enlisted Marine, he asked me why I had not visited the hotel in the past few months.

"Something doesn't seem right," said the sergeant major with a confused look on his face. "Why would they put you on hold, take a message and not call you back?"

We already sold Ball tickets to 250 people, including high-ranking officials out of D.C. I was able to work with a local military base to hire five daycare providers to take care of all the children. Every spouse was excited as many of them wouldn't attend an event where they couldn't have their children nearby. So, needless to say, this was huge.

I assured him that all would be fine, as they're probably just busy at the hotel.

"Let them know we are coming down tomorrow for a visit," the sergeant major said with authority. "Meet me here at 7 a.m."

"Aye, aye sergeant major," I said with a measure of fear and trepidation.

So I called the hotel, only to discover that the manager wouldn't take my call.

"We will let him know you are coming with the sergeant major," my point of contact said with a nervous sound in her voice.

The sergeant major was ready the next morning at 6:45 a.m.. We made our way to the Garden State Parkway and had a great conversation about his family, his background in church, what we believed about God and more. We were connecting in many ways—bonding through transparent conversation and hearty laughter—until we pulled up to the hotel.

An Unwelcome Surprise
"What in the world is going on," said the sergeant major as we both stared at bulldozers and what appeared to be a large, white circus tent blowing in the wind.

The building they promised was *not there!*

I felt a blood rush as if being emptied from my body. Confusion set in.

When we walked into the hotel, the folks at the front desk told us to wait in the lobby for what felt like the longest 15 minutes of my life.

"He better not tell me that our Ball will be in a tent," said the sergeant major, mumbling under his breath while pacing the floor. "This is not good…what are we waiting for?"

I was keeping my distance while praying. My career was done. My fitness report will be ruined, and I'll have to explain to my wife why there will never be another promotion. After all, I trusted the hotel management when they canceled each appointment leading up to the Ball. Nothing appeared out of line leading up to this moment.

The tension in the room was thick. Everyone involved was on edge.

"Sergeant major, the building wasn't completed in time, but I thought we could hold the event in an amazing tent," the manager explained. He was a tall, well-kept man around 60 years of age with gray, thinning hair.

"We've had events there, in…"

"You mean to tell me that all of the wives are going to be wearing their ball dresses in November outside in a tent!" the sergeant major exclaimed with a look on his face that made everyone in the room very uncomfortable. "What about my color guard?"

"They will be fine…"

"No, no, no, there is no way my color guard will be able to march into a tent," quipped the sergeant major, cutting off the hotel manager again. "Now I don't know what stunt you think you are pulling here, but we are not going to use this place for our Ball."

"Well, sergeant major, we are two weeks out," he said. "Everything is booked, and you really have no choice."

If there is one thing you never tell a sergeant major, never tell them, "You have no choice."

Big gulp, breathe, try to breathe. It gets "gooder and gooder" as my manager from Georgia used to tell me as a teenager working at a gas station.

"You have connections in Atlantic City. Call them!" said the sergeant major with a very calm but determined voice. "Don't you have contracts with the Pentagon. I know many of those folks. You don't want to tell me we don't have a choice."

"Sergeant major, I might know of a place that we could have the ball," the hotel manager said, writing on his note pad with illegible writing due to the extensive shaking of his hand. "Well, you call them."

They took all of us out to look at the "amazing tent" which made the situation worse.

An hour later we were grabbing a bite at a local restaurant, trying to resolve our dilemma.

"Sergeant Major, I know we'll figure this out," I said. He had been through many trials in his 25-plus year career, but nothing like this. I noticed he was staring at his food with a look of despair.

In the Corps, the Sergeant Major is the highest enlisted rank who is responsible for all the Marines in his area. When he speaks, we all listen and do as we are told. That rank puts fear in almost every Marine.

He was picking at his food, rubbing his forehead, making grunt sounds under his breath. The anger he displayed an hour ago in a tactful way towards a few civilians was slowly dissipating.

We drove back to Iselin, NJ, talking about our upbringing, family and faith with joy to a literal "fog of war" scenario that we wouldn't wish on our worst enemy.

A Novel Idea

"I will call the hotel we held the Ball in for years prior to this to see if they could help," I said.

A few minutes later I called from a pay phone and said…"Yes, this is Sgt. (LT) with RS New Jersey. Can I speak with John? Hi John, please forgive me for bothering you, but we are in trouble with our Ball venue here in Atlantic City."

"We completely understand," John said as he spent the next several minutes telling me how some of these deals can go wrong. "Call me back later today. We might have a solution," he added after hearing the entire story.

He said they had space for us, but it was only half the size as last year.

"Sergeant major, I called the Radisson and talked with John. He is very supportive and understands our situation," I said with a smile. "He is willing to meet with me tomorrow at 9 a.m. to show me what they have available. I know something good is coming."

The sergeant major came back to life after hearing this. I dropped him off at a local recruiting substation and he said,

"Call me at MEPS Philadelphia tomorrow if you think the event can happen at the Radisson."

The next day, the event pros handled me with care.

They knew what we were up against. The commanding officer was briefed, and my future was also on the line. The hotel had plenty of space to handle our ball. It would be tight, but we signed a nonbinding contract to have the Ball there with 250 guests.

John then gave me pointers on how to get what we wanted out of the hotel that left us hanging.

"Sergeant major, you can send your rounds down range!" I said from the hotel lobby. "We are locked in and have a venue now."

Two hours later, the sergeant major called me back and said "meet me tomorrow at the Taj Mahal! I told him if he doesn't lock us in to another place, that I will call my contacts at the Pentagon to let them know what he did, and we will cancel our ball."

The next morning was like watching the final scenes from the movie *A Few Good Men* only a little less dramatic.

The Taj Mahal manager was amazing. She could easily sit next to President Trump on "The Apprentice," intimidating anyone looking for approval to win the show. After a tour of the most amazing facility ever seen on this earth, we took our seats in a meeting room that I can only describe with an English accent as belonging on the show Lifestyles of the Rich and Famous.

"We can hold your event here in two weeks with the same meal choices," she said with little emotion. "Here is the price sheet."

"Well," said the sergeant major with a slight smile, "the hotel manager [Mark, from the previous hotel] will be paying the bill."

"Ok, well, the cost of reserving the room with meals is $20,000," she said, as Mark dipped his head and gasped for air.

"Hold on a second. You promised us that every spouse would have flowers," said the sergeant major with authority. "Add that up please."

"Oh, come on sergeant major," Mark said with a desperate plea to stop the bleeding.

"You also promised to have wine bottles for every table," the sergeant major continued with glee "can I see your wine selection please."

"Sergeant major!" Mark cried out in a loud voice. "This is too much."

"I can make a call to end this now if you would like," the sergeant major gave him a stern look.

This continued for 30 minutes. Mark contracted very expensive limousine buses to transport all of the Ball guests back and forth from the hotel to the Taj Mahal. He also reduced to the cost of the rooms and paid for the childcare and the food for the kids.

My career went from "no more promotions" to one of the best fitness reports in my life. I would be promoted to Staff Sergeant six months later.

That vision of having the Ball at the Taj Mahal came to life. The prayer reminded me of the following verse:

And whatsoever ye shall ask in my name, that will I do, that the Father may be glorified in the Son.— John 14:13

Watching all the wives exit the buses with their hand placed through their heroes' arm was a visual only captured in movies. I remembered those same wives were upset with me about their beef order two years prior. They were mad at the Corps for "stealing" their husbands. They were tired and weary from taking care of the kids, night and day with the father showing up after bedtime and leaving before the children woke up for school. This man who had sacrificed his life for our country seemed to never be recognized for how well he served.

That night the Marine Corps Ball at the Taj Mahal would put a pause to all of the pain. The spouses had time to drop off their children, head to the hairdresser, and prepare for a night to remember. The husbands who were weary from long hours of promoting love for the Corps were going to be alone with their wives for one special night.

Although our recruiting station had females in an administrative capacity, we didn't have female recruiters in our area at the time, so I only saw this from a perspective of what these wives appeared to live. This is my opinion.

They entered the hotel where their spouse was in full dress blues, clean shaven, in the best shape possible. You could see the excitement and hear them say "wow, look at those nice buses… are they taking us to the Taj Mahal?"

As they departed the buses and entered the casino, the patrons stopped, turned, stood up and clapped.

I could only imagine the spouses saying in their mind "This is my Marine. There are many like him, but this one is mine."

Tears flowing with each step walking as she walks into a room, greeted by an ice sculpture and a table with a glistening waterfall for drinks. The beautiful flowers greet her at a table made for a queen's banquet.

After the cake-cutting ceremony and speech were complete, the DJ started up; however, I was surprised to see the ballroom was nearly empty.

I walked out of the room to find out what was going on.

The sergeant major said, "follow me."

We walked to a giant room next door and saw one of the most amazing sights. A host of very rich folks had the Mike Tyson vs. Holyfield fight showing on a huge TV screen. They found out the Marines were there and brought them all in and bought them drinks.

We watched Tyson bite the ear off of Holyfield that night in what most would have to pay to see.

"Sgt. LT, you were right," the sergeant major said with a smile, "this all worked out for good."

The Taj Mahal was good!

CHAPTER SEVEN

The Good Walk!

And he said unto them, What manner of communications are these that ye have one to another, as ye walk, and are sad? And the one of them, whose name was Cleopas, answering said unto him, Art thou only a stranger in Jerusalem, and hast not known the things which are come to pass there in these days?
[Luke 24:17-18]

"One of the greatest parts of this job is that you can shoot all the weapons and head back to the office without having to clean them," I would tell my Public Affairs Marines. "You also get to fly in all types of aircraft and meet high-profile folks who many would love to spend five minutes with."

This public affairs job offered opportunities that most Marines wanted. The very first interview I had was in the office of a brigadier general on Parris Island. A few years prior to this, I was sitting on Okinawa inside of a green pop-up tent placed on two Humvees with a PRC-77 radio in front of me. We were told to work and not look at the commanding general as he passed by.

There is only one word to describe the feeling in your body when a general is watching you work—FEAR.

Walking into this interview, I had that same feeling.

Why send me? What was my new boss trying to prove? Any of the other Marines with experience could have gone.

The aide, a lower-ranking captain, brought me into the general's office for the interview, and the first thing the general asked me was "would you like a cup of coffee corporal?" as he poured a cup for himself.

This was a precursor to a what become a new normal in my life. If I would have never met my wife and reenlisted, the experiences leading up to the channel *And We Know* might have never occurred.

Fast forward to New Jersey about three years later and we would constantly run into high-ranking officials. Add to this job the responsibilities of working with the NBA on halftime events to promote the Marine Corps, MLB 7th-inning stretch pull-up contests and the support from the Marine Corps League. It was a like a dream come true.

One of the most important responsibilities of a Marine PAO was gathering information for a news article that would be sent to every base publication throughout the earth.

"Sergeant, we need you to head over to the border of Pennsylvania and New Jersey tomorrow," said the major from Headquarters Marine Corps PAO. "There is a former Marine walking across the state line with a Humvee escort. Get the story on him and send us some pics by the end of the week," he added before filling me in on the details.

Some "famous" guy started walking across the USA, from California, and was making his final trek to our state. Evidently, he was making headlines in every major city he walked into. The Marines caught on to his venture and agreed to have a Marine Humvee from the respective I&I area escort him.

I called the Pennsylvania I&I station to find out what time he would be crossing over so I could be there for the shot they were looking for—a man walking, with a hummer in the back and a sign saying, "Welcome to New Jersey."

After arriving about an hour before he crossed, I spent some time in prayer and I had a strong sense that there was more to this than just a guy walking a long distance.

As this "mystery guy" walked over the hill in my direction, I jumped out of my vehicle with my camera and ran up to him to make sure he was okay with me taking photos.

The walking man appeared to be his 50s. He was wearing professional gym gear with fine walking shoes. He had a very long hiking stick and a white bandana around his forehead which was filled with sweat pouring from his thick, black and grey curly hair.

Before I could get a word in, he shook my hand and said "God bless you my brother!"

"God has sent you for this venture. I can feel it," he added with a huge smile.

After several photos, he invited me to his next stop at a small hotel to grab some food and conduct an interview.

Why did he start walking? How long has his journey been? It turned out that this guy just had a strange feeling, walked out of his house in California and started walking. There was so much going on in his life. He needed a break.

Eventually, the Marine Corps found out he was a former sergeant in the Corps and somehow worked out a hummer escort on the freeway. The articles reached the highest-ranking members. The newspaper articles and TV news made it to the desk of a major shoe company which provided new shoes for the guy every 40 miles.

The strong, dark-skinned man worn from sun exposure, pulled out news articles from across the country. His reach to many across the Nation was huge—doing on foot back then what we do on social media today. Making contacts and growing by word of mouth.

As he shared the many stories from his journey, his face was glowing. He was closer to God than anyone I had ever seen.

He asked about my family, and he had a genuine concern for my wife who was not a believer in Christ at the time. The next thing you know he was in my car. Now, please don't ask me how a complete stranger ended up traveling with me to my house. I also can't recall how in the world he convinced my young wife to let him in. She was thrilled to make dinner and listen to his stories. Remember, this was before cell phones. I was able to take him back to where he left off his walking venture later that week.

The living room was filled with laughter as we talked about family, and he shared story after story with us. His Latino accent was very strong, especially at times when he was sharing something very exciting.

Eventually, he was led and somehow found a way to share the gospel with my wife. Now, later on in Iwakuni, she would be fully sold on the gospel, but this was a major turning point for us. The next thing you know they are praying together, and one week later she was baptized by him in a chapel baptismal on the military base where our housing was located. Well Glory!

Did it solve all of our problems? Were we living in ecstasy for the remainder of our lives? You know the answer to that, but I will say that there was an overwhelming softness that grew in our lives toward each other. More on that later.

After he finished the trek across America, we put a news story together for Marine publications and I thought it was done. Nope. He had a "great idea."

"Let's work with the city of New York for a final walk over the George Washington Bridge," he said looking at me as if to say "you can do this."

"There is no way that is going to happen," I replied knowing that that particular task would require a lot of manpower and hours.

"Let's pray about it," he answered, sitting in my office. He asked our Heavenly Father to help us put this together and that He would receive all the glory for it.

Now, all of this was happening at the same time that I had received orders to move to Iwakuni and my replacement was on the way in. This is where it gets really good.

Satan Enters the Building

Three months prior to the…let's call him the "walking sergeant"… entered my world, I received a strange phone call from my Marine Corps monitor. In the Army, they call them detailers. These are high-ranking enlisted folks who are responsible for the orders of each Marine in a specific job or Military Occupational Specialty (MOS).

These monitors hold the keys to your next 3-year duty station or base assignment.

The Good Walk!

"I have your e-mail here Staff Sgt. (LT)," he said with no emotion whatsoever. "It appears you want to go to Okinawa?"

"Yes, Master Sgt.," I responded as if great news was about to enter my life. Remember, the monitor put me in New Jersey instead of New Mexico three years prior.

"I am afraid we don't have any slots for Okinawa right now," he said with a sigh.

That is not what I wanted to hear, yet complete peace filled my heart and mind as he spoke.

Now, the crazy thing is that I just finished reading the Bible and these verses shortly before he called.

Trust in the LORD with all thine heart; and lean not unto thine own understanding. In all thy ways acknowledge him, and he shall direct thy paths.—Proverbs 3:5-6

The following words came out of my mouth, totally unexpected: "Master Sgt., I know that you did everything you could to get my family to Okinawa. I want to thank you on behalf of my wife for your efforts," I said with joy.

Say what? That is not what I promised my wife.

My wife told me to fight for those orders to Okinawa. Her friends, Marine wives, told her that you can get what you want if you fight. As a matter of fact, she was coached in the art of "Marine wife power" to get those orders. You would hear phrases like, "Call the sergeant major and complain if your husband doesn't get the orders you want. It works!"

The most amazing thing happened on that phone call. It would be the catalyst for *AND WE KNOW*®.

"Really, staff sergeant?" the monitor asked with a puzzled voice. "You aren't upset?" he added.

"No master sergeant," I replied. The words flowed from my voice and inner being as I realized that Holy Spirit was guiding my every word.

We briefly talked about his life and how he has to deal with yelling and anger almost every day. His job was not an easy one.

Then he said hold on a moment. He put me on hold for what

felt like an eternity.

"Hey staff sergeant, do you have any broadcasting experience?" he asked with a completely different tone of voice than when he first called. "I have a slot in Iwakuni that needs to be filled. It is an operations supervisor position at AFN, American Forces Network."

"I visit TV and radio stations every week and have made a few television stories for the Marine Corps," I replied, wondering if it was enough.

"That's good enough," he said with excitement. "I will have your orders for Iwakuni out today. Thank you staff sergeant for a great talk today," he answered with what I would call a pep-in-his-step, only it was a vocal pep.

"What just happened?" I thought to myself.

My next thought… My wife. Uh oh.

"Hey, I just got a call from the monitor," I said with trepidation. "And you know that God always…"

"Did you get Okinawa?" she said cutting me off. "Well, not really, but…"

"What? Why not?" she answered with a slight bit of anger.

"Listen, we did get overseas to Iwakuni though," I said with excitement.

"No, I don't want to go there," she said. Then she hung up.

The next day she called to apologize. Her mom called and said she had a dream that something was wrong in our household. Wow. Mom instincts or God using her mom to wake her up.

As my wife was complaining about my orders, her mom cut her off and said something like "that is your husband, and he deserves to be treated well for all he does for you and the boys. You need to get your act together and support him and get ready for your move. Love your husband!"

When I got home that day, she had luggage out. I thought she was packing to leave. Divorce?

She walked around the corner smiling.

"Are you ok?" I asked with fear in my voice and concern on my face.

"Yes, I am getting ready for our move to Iwakuni," she said with joy. What's wrong?" she asked.

"We aren't leaving now," I said. The joy of marriage, right?

Satan Enters

Ok, now that you understand the impact of the entire move we were about to make, this will make more sense.

So where were we?

1. The "walking sergeant" wanted to walk across The George Washington Bridge to NYC.

2. We are moving halfway around the world in one month.

3. A new sergeant is on her way to take over my position in New Jersey.

There was stirring in the hallway as I heard the new administrative gunny making her way towards my office. She was telling her that the public affairs NCO office was down the hall.

The "walking sergeant" was in a hotel a few blocks away.

We had just finished praying about the walk across the bridge.

"Hello staff sergeant. This is your replacement Sgt. (Zelda)," said the gunny.

After some chit chat, we started going over the turnover binder with all of the responsibilities for her new position.

"Why do you have a Bible on your desk?" she asked with a puzzled look. "That is not authorized."

This conversation came up once before as a gunny tried to tell me to remove it. After letting him know that it is authorized, he screamed at the top of his lungs and left. Later that day, we prayed together. Another book with that story will be forthcoming.

"That Bible is allowed," I answered her. "We have that right as Americans. And as Marines, we defend that right."

She stood up, reached into her blouse from the neck and pulled out a pentagram on a very long chain.

"I worship Satan," she said with a very strong demeanor.

"That's interesting," I replied.

It didn't faze me one bit.

"Let's get back to work," I said.

"You don't have a problem with that Staff Sgt. ?" she asked.

"Not at all. We all have our freedom of religion in this country. I worship the creator and you worship his creation."

She was astonished by the answer.

We continued our turnover work for the day. I told her about the "walking man" and she said we shouldn't be endorsing or helping someone using God to walk across the country.

The next thing you know our chaplain from the base called and somehow arranged to have the "walking Marine" walk over the bridge and meet folks on the other side. The date was set for the following week. I was in shock.

Even though Sgt. Zelda was not happy working on this project with me, she was stuck. She was new and needed to make a good impression to her boss, the XO.

The day of the big event came together in a magnificent way. We all met in Iselin, NJ that morning. I only had enough room for the "walking Marine" in my car. Sgt. Zelda had to ride with the chaplain. This was the same chaplain that I shared about earlier who received my tithe in an envelope because I didn't want to be tempted to spend it. We became close friends and golfing buddies, and he knew she worshiped Satan.

Ain't God good?

When we returned after several hours of work and traveling, Sgt. Zelda was in a somewhat sour mood. We prayed for her and asked our Heavenly Father to provide an opportunity for us to minister to her.

A few days later that prayer was answered.

"Staff sergeant, the people who were supposed to help me move my furniture are not coming," said Sgt. Zelda over the phone. "I also fell and have a cast on my hand."

Her very faint voice was also choking up with tears. She was a single mom who had her name changed recently so her abusive ex-husband wouldn't find her. Her daughter was about five years old. She packed her own household goods, put them in a U-Haul and traveled cross country to a foreign land—New Jersey.

The "walking Marine" happened to be in my office sharing another "great idea" that we will cover in the next chapter. Needless to say, he overheard our conversation.

"Hey, Sgt. Zelda, this is the walking Marine. I would love to help you move your furniture if you don't mind," he said with a very soft, loving voice. "Staff sergeant, LT, and I will come over tomorrow morning ok. Can you give us your address?"

She was beside herself. The tears were flowing. There was a hesitancy to approve this, but she said OK.

The next day we were moving furniture, the walking Marine was singing and filled with joy, and her daughter was laughing at all the jokes we were telling.

Sgt. Zelda's voice was also filled with happiness. This same Marine who was upset with all of the "God" things a week earlier was now receiving one of the fruits of the Spirit. LOVE.

After a long, hard day, we finished the move. The walking Marine sat down next to Sgt. Zelda and shared the gospel with her. I remember tears. I remember happiness.

How could God send a man across the United States on a whim? This man took all of his pain away by simply walking.

Each person he ran into was an act of divine Providence.

He knew that his Father in Heaven loved him. He was filled with the Holy Spirit. That gift from Jesus Christ is absolutely amazing.

I remember at that time how amazing it was to see God at work in a man who had a mustard seed of faith in believing Him. Not only that, he applied the Word of God. He didn't just study the Bible, sit in a pew and then forget it the next day.

He applied the Word of God in his life. *Literally.*

I will always remember that this man not only had a long walk across our Nation in a physical sense, he also had long walk with thousands across our nation, spiritually. Who knows the millions who will be changed by each step he took each day on his amazing journey.

I will always remember that *it was a very good walk.*

And it's not over, yet.

CHAPTER EIGHT

My Three Good Kings!

On his robe and on his thigh he has this name written:
King of kings and Lord of lords.
[Revelation 19:16]

After more than 20 years of active duty, I decided to put in for retirement. Details on how that all played out will be outlined at another time.

But the major who retired me insisted that we have a ceremony with a color guard and a formation of young Marines. She said, "This retirement event is more for the young Marines than it is for you."

Who would have ever thought that a man who reaches the age of 38 could say he is retired. When folks would tell me "that is too young to retire" I would often reply with something similar to the following statement.

"Did you have to move every three years? No. You were able to stay home and keep your contacts and grow your business and income. But we had to start over. Have you ever had someone call to say you have to leave in six hours without any indication of when you might be able to return? Have you had two hours to pack your bags, land in a military danger zone with a rifle without so much as a call home?"

Normally folks get the picture as I ramble on with more poignant questions. The stress of Marine Corps active duty is often intense.

There were many times I told my wife "I quit. I am not reenlisting!" Many of those moments would occur after experiencing some form of persecution for being a Christian in the Corps and sticking to the moral compass God gave me. I would be reminded of the fact that God's Word teaches that His people will be persecuted for their genuine faith and end up drawing strength to persevere from these precious Scripture passages.

In the world ye shall have tribulation, but be of good cheer; I have overcome the world.—John 16:33

Count it all joy when ye fall into various temptations, knowing this that the trying of your faith worketh patience. —James 1:2-3

So, I was going to get out after 17 years of active duty without retirement benefits, yet, something happened that changed my life and my decision.

I received a message from a KING.

Three years after that message, I was standing in a large theater with nearly 80 Marines in formation behind me while giving my retirement speech to tell them about that message.

The Marines in my office gave me a going away gift—a meme'd up "Windsock" newspaper with a headline "Master Sergeant gives longest retirement speech in history."

I titled that speech (sermon) "My Three Kings."

First King – General Krulak
"Staff sergeant, I was a sergeant for General Krulak when he was a captain in the Corps," said the walking Marine, as he was about to pitch another idea my way. "I heard that he will be in Atlantic City [NJ] next week. Why don't I walk one more time into that event."

You have got to be kidding me. He wants to pause the Marine Corps Scholarship Foundation golf event? This event had the Marine Corps Drum and Bugle Corps and some of the most distinguished guests in our nation.

On top of that, I am moving in two weeks.

"It's ok, staff sergeant," he said, as I told him I was moving. "Sgt. Zelda can hold the fort and you can work on this full time."

Somehow, he convinced me to send a letter to General Charles Krulak, the 31st Commandant of the Marine Corps, asking if the walking Marine could walk into the middle of the event with a Humvee escort as a finale to his long walking journey.

The commandant's public affairs officer called me to confirm our mailing address an hour later, stating that she received the letter and that the commandant read it.

Lo and behold, the word got out to many in our recruiting district, as well as our commanding officer and some at the Pentagon. They were not happy.

Folks, it felt like I hit a beehive.

There were Marine Corps officers from the district and from the Pentagon asking me to put a stop to the walk.

I was thinking, "Why are all these people calling me to ask about the walking Marine? And why are they telling me this idea is unprofessional?"

Upon hearing this news, the walking Marine would simply smile and pray, knowing that the enemy was hard at work trying to stop this walk.

A few days after we sent the request, the commandant had a personal letter sent to the walking Marine. He couldn't wait to meet him in Atlantic City. Gen. Krulak's staff would work with us to make it happen.

I had the privilege of meeting Gen. Krulak once before at a huge 3-on-3 basketball event in Northern New Jersey during my first year as a public affairs NCO. During such events, I worked with large companies to get booth space so the recruiters could meet young folks in their target market.

The commandant found out about the event and wanted to experience it for himself. When word got out that he would be there, all the district-level leadership got involved. They were my supervisors based on Long Island, NY. And these guys were sweating his visit.

The pressure was immense. It had to be done right, and they were determined to make sure that it would.

In order to understand how the commandant worked, I started looking for any type of videos we might have in our archive room. The public affairs office would send VCR tapes to each PAO office throughout our worldwide network with speeches, b-roll footage, and news clips for us to use when needed.

I happened to stumble across a videotape with Gen. Krulak giving a speech at a dinner somewhere in the United States. As I pushed play on that old, dusty VCR player, I felt a touch of the Holy Spirit prompting insight on my life as he spoke.

There was something different about this general. I didn't noticed anything at first—his uniform was impeccable, his smile was contagious, and he was short like me. I am only 5- foot 4-inches tall. But then, as he began to talk about the Corps with deep conviction, he paused and proceeded to share his "testimony." He was a believer in Christ!

"How can this be?" I thought to myself. "Is it ok to share your faith in Christ at a dinner like this?"

Tears were flowing as I realized the number one officer in the Marine Corps was not ashamed of the Gospel.

He was no ordinary man. This general shook the cobwebs off a stale, old Corps and started revamping it as soon as he took over. His faith in God not only had an impact on the Marine Corps, it also rippled through almost all branches of the military.

"How?" you might ask. Great question.

One of the first things he did was send a memo out to all Marines, called a MARADMIN (Marine Administrative Message), letting them know that he was setting up a panel to travel around the world to interview Marines to better understand was not working well and how the Marine Corps could improve.

Now for most civilians, this might seem like a wonderful idea. If you are a Marine stuck in the mindset that high-ranking leaders know everything, and who cares what the 18-year-old complains about…well, you would be in for what Marines call "a world of hurt."

"What would an 18-year-old know?" were questions often brought up at meetings when all of this was happening. "This will never work," many folks would say, mocking his decision.

Well, what do you know, young Marines started sharing how they felt about the current camouflage utilities, known as cammies, BDUs, battle dress uniform, or the fatigues—and how these were costing Marines too much money in replacement and dry-cleaning costs.

In those days, our cammies were made of a cheap material that would rip easily when out training in the woods or in the desert. Although we received an annual uniform allowance, those Marines who were constantly in the field training, would have to buy new cammies after nearly every deployment.

Another sore spot Marines had with these uniforms was one we faced regularly. We would put them in the dry cleaners and have them heavily starched so that we would look our very best for work and for wall-locker inspections. When you are spending $100 a month for dry-cleaning and $100 a month for new uniforms on a $900 monthly income, well, you can see how that can have an effect on performance and morale.

The review board would post these findings for all of us to see, and we were allowed to send complaints directly to the commandant's staff about our issues using a new system called e-mail. I believe it was called Marine mail.

As we would say in the South "Woo Lord." Gen. Krulak was turning over rocks and finding snakes everywhere.

Added to this complaint were the boots. The complaint came in that we're spending hours of our free time spit-shining the boots. They would get messed up in training, forcing those who wanted to be at the top of the game to buy new boots every few months to ensure their boots were perfect for every inspection.

After a few months of communication and research on these uniform problems, the Corps found a new and improved uniform for the Marines. I called them "pajamas." That is what they felt like. They were easy to clean, and we were told to no longer put starch on them.

They also abandoned the all-black combat boots. We started wearing the desert camouflage boots. Thankfully, no more spit-shining!

So, we knew who our commandant was. He was a Marine's Marine. We loved him and feared him, and we were grateful he was one of us.

Back to the 3-on-3 basketball event.
After setting up the entire event, the commandant drove up in what appeared to be a secret service car. We were freaking out.

"Here he comes," I said while scrambling to get my camera ready.

His uniform looked like a stiff board on him. I had never seen a more perfectly dressed Marine. He had a smile that was as bright as the sun.

Gen. Krulak brought all of the Marine recruiter participants together to deliver a short speech. He shared what they were working on to increase Marines' pay and benefits, how the recruiting numbers were looking from his perspective, and how proud he was of each Marine.

I will never forget how he finished. "God bless you, Marines!" he shouted. "Ooorah!" we shouted back.

As I walked up to get a few quotes from him for the story, the commandant, who was shorter than me, asked about my family and how the Corps was treating me. It felt like he really cared.

When we finished, he did something else that shocked me.

Gen. Krulak, using the palm of his hand, gave me a gut check and said "Semper Fi Marine. Keep up the good work."

Thank God I was a bodybuilder and had six-pack abs. Phew.

Atlantic City Event
The big event was upon us. The Marine Corps Scholarship Foundation really put on phenomenal events for attendees. Imagine the scene. You have a Marine Corps band playing, very well-dressed attendees, and decorations in the area filled with the splendor of the Marine Corps.

Attending an earlier event, I watched as the Foundation invited youth up to give them a post-graduate scholarship.

Their website, mcsf.org says "As our Nation's oldest and largest provider of needs-based scholarships to military children, the Marine Corps Scholarship Foundation helps cover the cost of attending post-high school, undergraduate, and career and technical education programs."

This organization didn't complain about the walking Marine at all. It was active-duty leadership.

The morning of the event, the sky was blue and the heat index was around 85 degrees..

This was the event so many high-and-mighty officers tried to have canceled. But as we had discovered, this Gen. Krulak was not like any other leader. He wrote a letter by hand to the walking Marine letting him know he remembered him as a sergeant in the Corps.

The walking Marine had a Marine Corps Humvee behind him during his walk to the event, and he appeared to have walked through a waterfall as he was pouring sweat from head to toe.

Each step he took was a step toward a finale that he would never forget. This moment was sealed in history long ago. The "walking sergeant" would meet the man in charge of the Corps. He never would have imagined this decades earlier when he first worked for the young "Capt. Krulak."

As the commandant was giving a speech to all the attendees, he paused as he heard the Humvee making its way toward the event.

The commandant, wearing his pristine uniform, stopped his speech, walked over and gave the walking Marine a huge hug.

The commandant's uniform was now dripping in sweat as he told everyone, "I want you to meet my friend." What a great moment for us who were there and it made us proud to be Marines.

Let nothing be done through strife or vainglory; but in lowliness of mind let each esteem others better than themselves.
—Philippians 2:3

Second King – Brigadier General Clifford Stanley

"Hey Sgt. (LT), did you know that a letter was sent to my office with your signature on it from a company?" said my public affairs Marine friend out of the Recruiting Station Atlanta office. "This isn't good, man. You could get in huge trouble for this as it's on the Marine Corps letterhead," he added with a voice of concern over the phone.

That call will forever be etched in my mind as I sat in a lonely, small office in the recruiting station. It came in right after a great victory where I secured more free advertising on a new TV station.

After he hung up, my phone rang again and I heard my old public affair's office commanding officer on the other line ask me, "Did you know you signed a letter endorsing a company?"

The lieutenant colonel was a huge fan of my work. He was now in the Pentagon with the "big wigs," and I knew that if he was calling me, then it couldn't be good news. He spent the next several minutes telling me that I could end up in the brig for breaking the law. He sounded sad and desperate—like he lost his son.

Just a few months earlier, I illegally used public affairs money to get all of the recruiters in the recruiting station brand new sweat gear with our logo on it. There was a lot of work involved in making sure the product was what everyone wanted.

The company asked if I could send them a letter to hang on their wall letting them know how valuable they were in helping me accomplish my mission. "Of course," I said. "I will fax it over to you today."

As most things in life, my work was important and based on my intros, readers know that the letter must be better than anything anyone has ever seen.

So, I got a copy of the Marine Corps letterhead from the main office and proceeded to type a nice one-page letter letting folks know that this was the greatest merch company on the earth. I signed it and faxed it.

They never told me that they were going to print that letter and mail it to every Marine Corps office on the entire earth— recruiting stations, bases, the Pentagon, U.S. Embassies and more.

E-mails were filling up my inbox all day about the letter they received. Fear crept inside my soul, though, as I didn't know the impact (if any) of this action. I actually called my wife and told her "I might not see you again."

I then informed the sergeant major and the commanding officer what was happening. They called the recruiting district CO, who was on his way to the office that week, to let him know what was going on.

They both told me that they had no clue what was going to happen but assured me that it would take time and that the brig was not on the table at that moment. They were confused. I was scared. It was a bad day.

The next day I received a call from the Marine Corps public affairs office that the new public affairs general, Brig. Gen. Clifford Stanley, would be at a university in New Jersey to give a speech to college students about the Marine Corps and that I needed to be there to get a story for a magazine.

That's strange, I thought. They didn't mention my letter at all.

Each day went by in slow motion. I would stare at the clock, waiting for 5 p.m., hoping no one would call me or let me know that someone was there to "take me in."

It was such a great feeling driving home each day, seeing my wife and two sons, knowing that one of these days could be my very last.

The General's Speech

Five days after the phone call about the endorsement letter my life was changed. I was on my way to the university in my Marine Corps Enlisted Service "A" uniform.

Entering the large auditorium, half-filled with college students awaiting for the general's appearance, I made my way to the top corner, hoping no one would see me.

My plan was to take some photos, write down a few of the things Gen. Stanley would say in his speech and escape.

The room was quiet as the announcer entered to introduce the first general in Marine Corps public affairs, Brig. Gen. Clifford Stanley. All PAO top officers prior to Stanley were colonels.

He was intimidating as he entered the stage wearing his Marine Corps Officer Dress Blue uniform. The Black Marine had to be about 6-feet 4-inches tall, and he looked like a bodybuilder, yet he was more than 50 years of age.

I only knew his age because he said he once achieved a perfect 300 physical fitness test score, but it was no longer possible once he reached the age of 50. Now folks, that means he ran 3 miles in under 18 minutes, was able to do 80 sit-ups in two minutes, and reached the minimum of 20 pull ups.

Then something I didn't expect happened. About ten minutes into his speech, he shifted his talk about the Corps and said, "I sing in the choir." He then proceeded to share that he had a normal life with certain hobbies, for the young college students to better understand, in my opinion, that he was human.

He then went quiet for a moment and I knew something different was about to happen. He started singing with the voice like an angel, "Amazing grace, how sweet the sound, that saved a wretch like me. I once was lost, but now I'm found, was blind but now I see."

I was in shock. This man had my attention. He shared that he was a Christian, that he loved serving his country and so much more. What a testimony, I thought.

Then he instructed us not to perceive that Christianity made him a timid or weak leader.

After he said those words, he yelled something like, "Hey, wake up!" It was so loud that the entire student body jumped, and you could hear a few screams of fear.

It was clear. Here was a man in uniform not ashamed of the Gospel. Here was a Marine not holding back on what true leaders are made of. I learned that day what true Honor, Courage and Commitment really looked like.

Completely forgetting my original escape plan, I ran down to the line of students to get an opportunity to meet the general and ask him some questions for the news story.

As I reached his aide, he asked me, "What is your name staff sergeant?" But I froze. The room suddenly went cold. My hands were shaking.

Everything felt like it was in slow-motion.

This aide was letting the general know each person's name in advance of shaking their hand.

As I came to my senses a few moments later, I said in a very fast voice, "Staff sergeant *(^&*." That's right. I spoke so fast; I didn't want the general to hear my name for fear that he would know I was the one who had endorsed the merchandise company and sent the letter.

The general looked at me and asked me my name. As I told him, he looked into the sky and said, "Wait a minute, how do I know that name?"

The aide opened his big mouth and said, "This must be the Marine who wrote that [endorsement] letter for the company, sir."

My heart was beating so hard, I thought it would explode.

"So, you're the Marine who wrote that letter. That's right," he said with a huge smile while shaking my hand.

"Yeeessss sir," slowly made its way out of my brain to my mouth.

"Staff sergeant, I read that letter and thought it was one of the best I have ever seen," he said with a joyful tone. "Yes."

As a matter of fact, you made me believe in that company. I would have bought their gear based on your letter," he continued with that amazing smile.

All of the sudden, my head looked directly at him with pride. This was not what was expected.

"Don't do it again!" he said with authority. "Now what were you going to ask me?"

"I won't sir. I promise. Thank you, sir. No questions sir. I want to thank you for an amazing speech," I said quickly as I looked for the nearest exit.

The first thing I did was find a pay phone, call my wife and tell her, "Guess what, I am not going to the brig."

Stanley served 33 years in the Corps and retired as a major general, eventually becoming the Undersecretary of Defense for Personnel and Readiness.

Whosoever therefore shall confess me before men, him will I confess also before My Father which is in heaven. But whosoever shall deny me before men, him will I also deny before My Father which is in heaven.—Matthew 10:32-33

Third King – Brig. Gen. Mastin Robeson

After reaching 17 years in the Marine Corps in 2005, I was through with the Marine Corps. My plan was to leave without reenlisting, forgetting all of the benefits that come with a 20-year military career retirement check.

Each decision I made that had moral implications would cause tremors in the office. The master gunny in charge of the office in Okinawa put me through hell, a young officer was under investigation, we just finished a real-world operation in Thailand after a tsunami killed more than 250,000 people in that area, and another officer proceeded to scream at me for teaching the Marines in the office that using profanity was illegal under the Uniform Code of Military Justice.

I would tell the Marines that if they had true Honor, Courage and Commitment, they would not to break the UCMJ, and have the courage not to curse like many of their peers and the commitment to stick to it.

Now, don't get me wrong. You might assume I was an "all law" master sergeant. I was simply establishing a precedent and would provide some latitude for many things, including occasional cursing.

As a matter of fact, Marines would sometimes ask permission to curse, in my presence, and I would let them get it out. They would then say, "Thank you" and move on.

General Robeson played a key role in keeping me strong to my 20th year.

We were in my final exercise called Balikatan in the Philippines. My job was senior enlisted public affairs chief for the joint public affairs operations center.

The work leading up to this exercise was immense. I was responsible for ensuring that all public affairs specialists from each branch of the military arrived safely to their destinations, and that they were set up with a room, money in their bank account, and more.

Added to this was my responsibility to vouch for the completion of their assigned tasks on time each day.

I was required to give a brief in the Joint Operations Center to all of the high-ranking officers each day as our one public affairs officer, beginning when they arrived a few days after the exercise kicked off.

These were not ordinary briefings. I made the journalists put presentations together similar to my *And We Know* broadcasts. In these presentations, I would normally play a 60 second radio news clip, show several photos we released to the media and discuss how their efforts and communication with our journalists helped us get what we needed to let the world know about their hard work.

As we were about to start the exercise, a massive rainstorm hit one of the areas of the Philippines that caused a mountain to fall onto a village, covering a school filled with more than 200 children and teachers.

Now, imagine the shock in finding out that you have to cover a real-world event and continue the exercise at the same time.

The captain I worked for said he was going to head down to the village with the 31st Marine Expeditionary Unit to work with the media as they covered the story of the military working to dig those people out.

Double the work. No sleep. No one to take my place for breaks. And then another call came in after the captain's call.

"Hey brother," said chaplain Biadog. "Could you preach for me Sunday at the local chapel near Clark Air Force Base? I am heading down with the 31st MEU."

"Of course," I said. Not wanting to turn down an opportunity to preach.

Now, remember, I was ready to get out of the Corps and this was "my last exercise and we're done."

Then a call came in from Okinawa.

"Master sergeant. I need you to have a speech ready for Brig. Gen. Robeson," said the public affairs captain. "He will be arriving tomorrow to speak to the press."

I had never written a speech in my entire life. Neither had any of my journalists.

I started looking for background information on the general. The best place to find information would be his website. He was the 3rd Marine Division Commanding General.

On the very front of the website, the general wrote that the 3rd Marine Division believed in the following three things in order: God, Family and Corps. It proceeded to say that if anyone had questions about this that he had an open-door policy for those who wanted to talk.

"What? You can do this on a government website?" I asked myself, as the tears started flowing.

I clicked on his email link and sent him a direct message and apologized for it being personal, but wanted him to know that I was struggling with being a Christian while being in the Corps.

He wrote me back with the most amazing letter about our core values and encouraged me to stay the course and included a few Scripture verses.

I was not only planning to reenlist now, but was also excited to preach on Sunday. I made it to the chapel but was barely functioning on three days with no sleep. My lance corporal wanted to watch. If it wasn't for him telling me what time it was, we would have missed it.

We read the story of David and Mephibosheth and cried together during the sermon. I was weak, but the Lord's strength was my strength.

The general arrived the next day. I assumed we would hang out and talk. He was all business, though. We showed him the podium, he shook my hand and said, "thank you." The press asked him some questions and I departed knowing it was all for good.

My Retirement Speech

I told everyone who attended my retirement on Jan. 31, 2009, about these three kings and their influence on my life. It was a long speech, but it was cleansing. That speech was not rehearsed. It all became clear in my mind just the day before. I simply wrote their names down on a piece of paper and shared the stories.

At the end, something came out of my mouth that was not on paper.

"These three kings changed my life forever. They all serve the King of Kings, and his name is not a curse word," I said with authority as I stared at the sergeant major who had a habit of using Jesus' name in vain.

"His name is Jesus Christ."

Thinking that it was going to cause issues with the Marines there, I tried to make a beeline out of the auditorium.

"Excuse me master sergeant," I heard faintly. Stopping, I turned around and there was a young Marine with tears in his eyes.

"I just wanted to tell you that I plan on staying in the Corps for 20 years faithfully as a Christian," he said while shaking my hand. As he talked, I noticed there was a line of young Marines forming who were waiting to talk to me.

The next young Marine shook my hand, her lips were quivering and tears were streaming down her eyes. She couldn't talk.

"It's ok. I understand. God bless you," I said as she walked away.

"Master sergeant, I am heading back to church thanks to you," said the young Marine. He shared how his mother had been begging him to live for the Lord, and how my speech inspired him.

One after another shook my hand and said, "thank you," and shared a thought or a bit about their own spiritual journey.

Two weeks later we were on a plane for Germany with our five children to begin our next great adventure of the *And We Know* family.

We will get to that later. For now, I want to leave you with this word of encouragement.

Don't ever think that your life is a mistake. Every second we have on this earth is a blessing.

While on this earth, never be ashamed of the Gospel.

And he said to them all, if any man will come after me, let him deny himself, and take up his cross daily, and follow me. For whosoever will save his life shall lose it: but whosoever will lose his life for my sake, the same shall save it. For what is a man advantaged, if he gains the whole world, and lose himself, or be cast away? For whosoever shall be ashamed of me and of my words, of him shall the Son of man be ashamed, when he shall come in his own glory, and in his Father's, and of the holy angels.—Luke 9:23-26

Chapter Nine

Cursing, Gothic, Satanists = Good Harvest?

And Jesus went about all the cities and villages, teaching in their synagogues, and preaching the gospel of the kingdom, and healing every sickness and every disease among the people.
But when he saw the multitudes, he was moved with compassion on them, because they fainted, and were scattered abroad, as sheep having no shepherd. Then saith he unto his disciples, The harvest truly is plenteous, but the laborers are few; Pray ye therefore the Lord of the harvest, that he will send forth laborers into his harvest.
[Matthew 9:35-38]

In the year 2000, I was stationed in Norfolk, VA. As a reminder, whenever Christians in the military receives orders, it is exciting because we know the Lord is guiding us into a new adventure where we will have an opportunity to reflect Him in our lives, knowing that our faith will be tested.

My job as a gunnery sergeant was to run a press department responsible for delivering nearly one million news releases to every sailor and marine's hometown. If a Marine was promoted, they would fill out a hometown news release form and send it to our department. We would then place that person's name in our database, run it through our system, process the release and off it went to their hometown newspaper, radio or TV station.

This was a very unique responsibility for an E-7 in public affairs, to say the least. During my first week, I interviewed all of the sailors and Marines who worked in my department. We had a few public affairs specialists, but the office was mainly filled with sailors who were being processed out of the Navy or going through some type of hardship.

Those sailors facing hardship needed a job that was not stressful; therefore, orders were granted to move away from the ships and to our shore duty assignment. Their job was to simply input all of the information from the forms into a database program.

Imagine sitting there, day after day, watching your desk fill up with envelopes. Each time you finished a stack, another box of envelopes would be dropped off. It reminded me of the 1980's Dunkin Donuts commercial where Fred Baker would say over and over "Time to make the Donuts" in rain, snow or shine at wee hours in the morning.

As I was interviewing my journalist 1st Class sailor (JO1), she proceeded to tell me of the many "characters" who worked in this strange office.

"Gunny, we have a pregnancy, a sailor facing a court case, a Marine who was caught with 'male-to-male' porn on his computer, and a sailor who was getting out due to stress," she said.

It became apparent that we had a gang of misfits. Each one had a different story to tell. To my amazement, the same week I checked in, they sent a sailor to my department, to what I started calling the "Psych Ward."

"Excuse me gunny," said a small, quiet voice after a knock on my office door. "I am here to check in," said a man who appeared as though he had just come from a Halloween party.

He was wearing a long, black trench coat with a black shirt and black pants. He had on black mascara over his red eyes. To add to the drama, he handed me his orders and I was shocked to see his fingernails were painted black.

"Is that authorized?" I asked, referring to his fingernails. He nodded with a yes and didn't say a word after that. I know the Scripture says to "Fear Not," but that was one time I was actually in fear. He made me feel very uncomfortable.

So, after some praying and reading the Bible, I was determined to treat everyone with dignity and make sure we put out great work each day. I knew that our Heavenly Father would reveal someday why this was my appointed place of duty.

Satan Worshiper

"Gunny, can I ask you for a favor," asked a young, bald sailor, Perez, who appeared to be filled with anger. "My car broke down and someone said that you live near me. So can I get a ride to work," he asked with some hesitation as he stared at the Bible on my desk.

As I stared into his dark eyes, looked at his posture and determined that there might be some demonic activity going on with him, I started to say no; however, the words poured out with a different sound, "Yes."

Since he would be in my car, he would have to listen to my cassette tapes, I thought. My car was filled with tapes of sermons from pastors and teachers of the Bible. For some reason, I sensed it would be good for him to listen to apologetic material from Bible teachers who would visit universities around the world. Apologetics comes from the Greek word Apologea, which essentially means to give an answer in reply.

As I pulled into his run-down, dark apartment complex, to pick him up, he ran towards my car with a hoodie over his face. It freaked me out. Perez opened the door quickly and jumped in. Not a word. No "Good morning." The young sailor simply put his hands in his hoodie jacket, pulled the hood over this head and looked down as if he wanted to sleep.

Pushing play on the tape, I proceeded to drive and knew it was going to be a long week. The sermon was filled with information about Nietzsche and Darwin, topics that I was new to.

The teacher was a master at bringing these topics together with Scripture and stories to paint a wonderful, biblical picture related to the times we are living in. He was able to show the lies from the world's teaching in contrast to the offer of eternal salvation of Jesus Christ that is the only solution to fill the emptiness these philosophers, Nietzsche and Darwin, were offering.

During the 35-minute drive on that cold morning, I was filled with a burning heart. I knew that a seed was planted although Perez barely moved. As we pulled in to his work, he thanked me for the ride and ran for the office building. He did the same thing as we drove home. Not a word. Though, he was forced to listen to the sermon I had playing.

On the second morning, we were halfway to the office when he spoke.

"Who is this?" he asked. "How does he know so much about Nietzsche? I like this, gunny," he said as he removed the hoodie from his face.

That started a dialogue that continued throughout the week. He started smiling, asking questions each day as we drove to work and back.

At the same time this was occurring, I was going through Evangelism Explosion (EE) training at our wonderful church in Newport News, VA. We prayed consistently for this man I was driving back and forth from his job each day.

On the final day of our time together, Perez invited me into his apartment. As I entered the door, you could feel a cold, dark force. It was right out of a horror film with scary music. As I walked upstairs, there was a white goat's skull on the wall. Turning towards the living room, you could see feces on the wall, food all over the kitchen floor, lights were turned off and their little girl was running around naked.

He then proceeded to tell me that he and his wife, Jane, worshiped Satan. Not what I expected to hear, though I was not surprised based on what I saw in his apartment.

"Gunny, you have changed my outlook on life this week," he said with a soft tone and glazed eyes. "Can you tell me how my life can be changed to be more like yours?" he asked.

For some reason, it seemed necessary to delay the evangelism part. "I will let you know as we drive to work tomorrow, ok?" I responsed. Perez agreed.

The very next morning, we pulled over and Perez prayed to receive Christ. How do I know it was real?

Two days later, we were invited to visit his apartment. The pastor and I pulled into the parking lot and prayed. As we walked to the door, I was expecting a "demon" to fling open the door and chase us away. That place gave me the chills.

Perez insisted that all would be fine. He wanted the Pastor and I to visit his wife. She was not happy.

"What did you do to my husband," said Jane, as she held a fork in her hand in a way that appeared threatening. "He stopped yelling at me and he seems much different."

Now, that is not what we expected to hear. Normally, one would assume that was good news, but she didn't.

I was able to share the gospel with her and then I left.

A week later, Perez, walked into my office during lunch with a huge smile on his face and asked if he could use my phone.

"Hey Jane, I'm in gunny's office now," he said handing me the phone.

"Hello," I said wondering what was going on.

"Gunny, I did it. I prayed," Jane said with the soft voice, much different than the last time we talked. "Can you come over today after work with my husband?"

Entering their apartment was much different this time around. The smell of flowers filled the air. Their daughter had clothes on, the kitchen was clean, the skull was gone, and the curtains were open, letting the sunshine in.

As they played with their daughter and shared their story, it was apparent that there was a transformation.

Leaning down to their toddler, I started sharing the story of Noah's ark. The most amazing thing happened.

I looked up at the wife and she had tears flowing from her eyes. They indeed had an encounter with God and a new, hope-filled life ahead.

First Bible Study at the Corps

Over time, many people started noticing a shift in the "spirit" of our workspace. Folks were coming to my office during lunch to share their problems as if I was the chaplain.

Because of this, I visited the chaplain's office on the base and let him know I was studying to become a pastor and asked permission to hold a monitored Bible study in my office every week. Several folks in the building were stopping into my office to talk to me about their problems and for some reason think I can bring them some answers.

As lives started changing, my curiosity peaked. That led to a rapper's visit, the homosexual's visit and the gothic's questions.

The Gothic

The gothic, who wore only black and loved his black fingernails, was about to go on leave for two weeks to his hometown in Georgia. As I signed his paperwork, he asked if we could talk before he departed.

"Gunny, do you have anything I can listen to while I drive home?" Jones asked, anticipating a positive answer.

"Well of course I do, Jones," I replied. "There is a pastor from Georgia that I listen to almost every day. I just got his tapes in the mail. You can borrow them, but please don't lose them."

He was smiling, which was not a common thing from that 21-year-old sailor. Before this, he never showed emotion or talked to anyone in the office.

We headed out to my car, and I gave him a new tape series on the book of Malachi from Johnny Hunt, pastor of First Baptist Church, Woodstock.

Johnny Hunt touched my life in Virginia at a small men's conference he was invited to speak at. This man was short like me, loved the Lord and he could tell stories that filled your heart with a burning that we read about in the Bible, when Jesus approached the couple on the road to Emmaus (Luke 24).

As a matter of fact, he was a master at applying faith in God to many who would typically think Christians were crazy. I remember thinking, "Lord, I want to be like Johnny Hunt…filled with joy telling stories of how amazing You are."

My favorite story was how Johnny Hunt invited all of the coaches from the high schools in Woodstock when he had just become a new pastor in the area. They showed up in mass and he told the coaches that they prepared the young men physically, but that he would like to prepare them spiritually.

He invited all the coaches and their teams for a barbecue at the church. Now, it didn't take long for the other sports teams to find out about the picnic and were looking to be invited too.

Well, the problem was that our small church would have to pay for hundreds of meals as almost every team from the high schools in the area confirmed that were coming. The church didn't have the money, but Johnny didn't sweat it. He knew the Lord would provide. Coaches were handing him checks, and donations were coming in from out of nowhere and the church was able to easily pay for all the meals and Johnny Hunt was able to give his message to all of them.

Well, imagine how the gothic, Jones, felt listening to story after story from Johnny Hunt about how faithful the Lord had been on his journey. Johnny Hunt loves sharing his testimony about how the Lord saved him and how he was transformed from the pool room to the pulpit. He literally led almost everyone in his sphere of influence to Christ after his conversion.

Two weeks later, I heard folks laughing and carrying on in the office. My curiosity got the better of me, so I slowly walked out of my office to see what all the commotion was about and I saw Jones talking to everyone.

He was wearing green Duck Head shorts and a blue Polo shirt with saddle shoes. All of the black was gone! In addition, he seemed to be "shining" in color as if he was with Moses on Mount Sinai.

"Gunny, thank you for the tapes," he said walking into my office, noticing the shock in my face. "I was listening to Johnny Hunt in the car and after two hours, I had to pull over because my eyes were filled with tears and I couldn't see the road!"

"I prayed right there on the freeway, gunny," he said with a calmness only seen in a person filled with peace. "And then a strong sense came over me that I needed to go to my home church."

"So, I went to my church that Sunday and was dressed up nice, so they let me sit up front," he said. "They never did this when I was wearing all black. They always forced me to stay in the back, even though I wanted to sit up front with my family."

He despised them for this. They were treating him poorly as a gothic.

"Then they asked me to share my testimony, gunny," he said. "This is where it gets good. I got up there and chewed them out."

"I told them that they were all hypocrites for treating me so nice since I was wearing nice clothes," he exclaimed as his voice started getting much louder. "You always put me in the back and never showed me love, but today you love me because I look normal to you."

He then took them to the following Scripture:
For if there come unto your assembly a man with a gold ring, in goodly apparel, and there come in also a poor man in vile raiment; And ye have respect to him that weareth the gay clothing, and say unto him, Sit thou here in a good place; and say to the poor, Stand thou there, or sit here under my footstool: Are ye not then partial in yourselves, and are become judges of evil thoughts?

Hearken, my beloved brethren, Hath not God chosen the poor of this world rich in faith, and heirs of the kingdom which he hath promised to them that love him? But ye have despised the poor. Do not rich men oppress you, and draw you before the judgment seats? Do not they blaspheme that worthy name by the which ye are called?—James 2:2-7

He walked out of the church, got in his car and left. He told me, "it was the greatest thing he had ever done in his life."

As my wonderful Pastor Abbott would say, "Well Glory."

The Rapper
"Gunny, we need to talk," said the tall, black man, as he slammed my office door closed. He then walked over to my couch andsat down , staring at my Bible on my desk. "I need help. Can you tell me what the Bible says about my situation?"

"Now, Parker, you know I don't have discussions like this unless it is lunch time or after work," I said. This was my rule so that no one would accuse of me sharing Christ during work hours.

Tears started rolling down his eyes. I sensed that this was real. He needed help.

"Gunny, I gotta baby and a girl in D.C., and it is tearing me up," he exclaimed. "My whole life is falling apart. I don't have a job lined up when they kick me out of the Navy. You seem to have the answer for all these other people. How about me?"

Phew. Take a breath. Pray. Listen quickly and speak slowly.

I went around to the couch and sat next to him. He poured out his entire life story to me in tears. He was shot, had been in some gang-related activities, was trying to express his burdens through his rap music, and was confused about the purpose for his life.

The gospel poured out of my inner being. I read the Scriptures to him, asked him some revealing questions and prayed for him.

When he left, I can honestly say this is what I thought "I don't think he heard a word I said. He will never change."

The next morning, I came to see how wrong I was!

"Gunny, gunny, you won't believe what happened," said Parker as he ran into my office and paced back and forth. "I got that thing. I got that thing you was talking 'bout—Born again."

"How?" I asked. "There is no way, please leave my office. We will discuss this later," I said, pointing to the rules again.

He shut my door again, looked at me with a glow in his eyes and said, "you have to hear this."

"I went to my room and got ready to go to the movies to see a horror movie," he said, still pacing. "Everything in my mind was freaking me out, gunny. I knew what you were saying was true, so I got on my knees and prayed."

"Ok Parker, many folks do this, but that doesn't mean you are born again. Give it time and see if you bear fruit, ok," I said, still doubting it, believing I was being deceived.

You are probably wondering why I was so tough and doubted him. Well, after many years of watching people pray, I witnessed many going back to their old ways, and it would always break my heart. It reminded me of the Parable of the Sower that Jesus shared with all of us (Matthew 13:1-23). So many folks would get excited, but the world would choke them and they would go back to their old ways.

"Gunny, I got into the car with my boy, he put my rap CD in the player, and I ejected it," he said with a desperate tone in his voice. "When I started hearing the cussing, it made me hurt inside, so I took it out and my boy asked me if something was wrong."

"Then we got to the movie, gunny. I love scary movies," he said again, trying hard to convince me something happened. "But this time I couldn't watch it, so I kept getting up and leaving and coming back."

"So, that doesn't prove anything," I said.

"Hold up, I ain't done," he exclaimed with authority. "When we left the movies, I was staring down this random brother as he was walking at me on the street and I looked down," he said, looking at me for affirmation.

"So," I said. "What does that mean?" He was frustrated cause I didn't get it.

Parker then proceeded to tell me that when you grow up in the streets of D.C., you have to stare people down as you walk past them to show that you can take them. He said it has to be mean look, so you teach yourself to look as if you hate that person.

For the first time in his life, he was staring a man down and looked down. It was such a big deal to him, he started shouting "Hallelujah, gunny—I got born again."

I called his friend downstairs who used to worship Satan. He ran upstairs, opened my office door, saw Parker and they started high-fiving and hugging. They were both free men—free from the bondage of sin. Their past was gone.

Tears flowed. It was a miracle.

The Homosexual

As each day progressed in the office and lives were being changed, there was one Marine that I knew needed help. He sat in the corner, didn't talk to anyone, was about 30 pounds overweight, and was suicidal.

He had a picture of his supposed "wife" on his desk, but everyone knew based on her look and the fact that he was caught with pictures of him with other men in bed, that he preferred the "alternate lifestyle." In those days, "Don't ask, don't tell" was the military law.

Nowadays, thanks to "O" (our former president, whose name I refuse to say, so as to not corrupt this book), they can now marry and have adopted children, live in base housing and more. If I was still in the Corps, we would see this as an opportunity to discover how God can work to change their "want-to's," as Johnny Hunt would say.

Well, after much prayer, the Lord led me to give him a project. Each day, I would tell him what a great job he was doing. He would complete a project and I would simply give him another one, congratulating him each time.

So, he was essentially exposed to a side of Christianity where the Christian didn't judge him. He knew lives were changing. He was watching this happen to others in the office, one-by-one.

Imagine how he felt when seeing signs on churches that said, "Homos are going to hell." I remember seeing signs like this all over the place growing up. The churches didn't really teach us how to help them, but we knew something was wrong.

One of my high school teachers was exposed and subsequently killed himself in Albany, GA. The embarrassment of being known was too much for him to handle. Who knows how it might have turned out for him, if a Christian church tried to comfort and guide him when his story was plastered all over the newspaper.

Again, the Bible really convicted me. Jesus was accused of hanging out with the drunkards, the tax collectors and sinners of His day—by the "church folk."

I knew that the Holy Spirit was guiding me to simply be kind and to make sure he knew I was not judging him.

Well, one year later, the young Marine walked into my office asking for help for his cousin who was "gay."

"Gunny, what does the Bible say about my gay cousin?" he

asked with much curiosity. "I am driving home tomorrow and figured he needed help."

Now, I knew he didn't have a cousin. He knew that I knew he didn't have a cousin. It was all about him.

I grabbed my Bible, took him to Romans 1 and started sharing from the Scriptures about sin—that we are all sinners. I told him that I was also guilty of sin, and that his cousin was the same as me. And we all need a Savior—Jesus.

He was overwhelmed as the Gospel was being shared.

You could see his face light up as he came to understand that each Scripture verse was there for him.

That Marine showed up two weeks later with a smile never seen before.

He walked into my office and said, "Gunny, my mom and I got saved."

"We had a car accident and ended up in the hospital, so I started sharing all that you said," he continued with a smile. "It is the most amazing thing. We read the Bible and saw how amazing it is."

Within eight months, that Marine lost 30 pounds, was on a meritorious board for promotion, rose to sergeant, and was accepted to recruiting duty.

Did everyone accept Christ in that office? No.

Were there times that I tried to win others to Christ throughout that entire three years? You betcha.

Overall, that time in Virginia proved to be a *good* harvest!

Then saith he unto his disciples, 'The harvest truly is plenteous, but the laborers are few; Pray ye therefore the Lord of the harvest, that he will send forth laborers into his harvest.'—Matthew 9:37

CHAPTER TEN

God's Good Hand in Broadcasting

And whatever you do, do it heartily, as to the Lord and not to men, knowing that from the Lord you will receive the reward of the inheritance; for you serve the Lord Christ.
[Colossians 3:23-24]

As I mentioned in previous chapters of this autobiography, our Heavenly Father found, in His great wisdom, to ensure each assignment in the Marine Corps and government service prepared me for the future—*And We Know.*

My recruiter, you recall, said welcome to the Corps with the opportunity to fly. I loved airplanes. It turned out that Aerial Navigation was not on the table for an 18-year-old at the time. They sent me to Air Support Operations to become an operator.

Although this MOS (occupation) had the word "Air" in it, the only time we were around aircraft was when they flew us to our next location in the mountains, or a CH-53 dropped off a generator in the field of Korea for an exercise.

We simply turned on our PRC-77 radio, used a sheet of paper and talked to a Marine forward observer who was calling out coordinates for a target that needed to be pummeled by a military jet or coordinates for a helo (helicopter) medical evacuation (MEDEVAC).

The common phrase used in that communication was "Say again, over!" You see, these radios didn't seem to work very well. They were connected to wires that our communication Marines would run from a small antenna at the top of a mountain nearby. I often felt sorry for these Marines, who were constantly walking up and down those mountains ensuring we had clear comms. The fact that they worked so hard had a lasting impact on me and would lead to a crucial support network I would need years later in Thailand during a Tsunami in 2004.

I was then moved to a different MOS called Public Affairs where I gathered information for "The Boot" newspaper. They never paid to send me to DINFOS for training due to lack of funding, so I simply studied local newspapers and learned the patter for writing. Six months after entering the new world of Public Affairs, I was awarded the MOS.

While in that office, I fell in love with Quickdraw, a computer program that enabled me to create graphics. Right next to the computer there was a book that provided a step-by-step tutorial on how to use the program, which is a similar, more basic version of what Adobe Illustrator is today.

"We need to let the folks on the base know that there will be construction on the parade deck here," said the staff sergeant. "We need to find a way to get this information to everyone to get them excited about it," he continued.

I jumped on that story, worked for a week on a graphic design showcasing the future parade deck. In order to get it approved, they sent me over to the engineering department..

The commanding officer said my design was "astonishing" and not only approved it, but he also asked how they could bring me into their MOS to do more design work for them. But he was disappointed to find out that I had just moved into my current job.

That particular graphic won awards in our annual Thomas Jefferson award program and was placed in the *Leatherneck Magazine*[3]. This was an honor for someone who had never attended public affairs school in DINFOS.

Leatherneck is a nickname given to Marines. It is derived

---

3 Leatherneck was an official Marine Corps publication until 1972, staffed primarily by active-duty Marines. In 1976, the Leatherneck Association merged with the Marine Corps Association (MCA). As of 2016, MCA continues to publish Leatherneck alongside another Marine Corps periodical, the Marine Corps Gazette.

from the leather neck-piece that was part of the uniform of the Marines from 1798-1872.

On a different and more personal note, I was also preparing for my very first bodybuilding competition. I was serious and knew this had to be done right. I bought a book that had valuable instructions from champion bodybuilder, Arnold Schwarzenegger, and I also saw Franco Colombo. He was a 5-foot 4-inch man who competed in the Mr. Olympia competitions. I decided it had to be done his way since we were the same height.

I realized early on that for me to be competitive, I needed to get rid of the hair on my chest. Instead of simply buying clippers and going to town, I had to be like Franco—just rip it out. Franco would use his bare hands and pull. Well, I knew that wasn't going to work for me, so my wife and I came up with an amazing idea to buy a cheap hair-removal wax kit from the store.

She heated up the wax on the stove and used the wooden spoon provided to pour that hot wax on my chest. She would let it get somewhat hardened and then would say "Ok, here we go, 1–2–3…" she would say squinting at my chest and pulling super-fast to minimize the pain.

"Ahhhhh, that hurts!" I would bellow out as she did this for 6 hours straight. Sometimes it appeared as if she enjoyed watching me in pain. No pain, no gain, right?

A week later was my first competition and they handed me the trophy for winning "Mr. Beaufort." That headline appeared in our newspaper and caused a buzz on the base—folks were pretty excited. As a matter of fact, the general's staff anxiously awaited my arrival for an interview they set up the following week.

When I walked in, the secretary saw me and said, "wow, you are a lot smaller than I expected."

"Well, ma'am, the cuts in the muscle with the stage lighting make the body appear to be much bigger," I tried to explain.

After filling the room with a heavy sigh and without apology, she walked away.

I was wired for competition. I couldn't do things halfway. It was either put everything in my mind to a given task or don't try at all.

In fact, my life was filled with competition; even going back to my younger days with the pinewood derby competitions in the Scouts, Bible-drills, math drill meets, baseball and so much more. One thing that would always drive me crazy—not winning.

Broadcasting

So, imagine my surprise when they gave me the orders to American Forces Network in Iwakuni, Japan. I had very little experience in that world. Remember, I just completed marketing in New Jersey, Philadelphia and some parts of New York. All of this was only a few years after being accepted into a job in Public Affairs.

Instead of sending me straight to Iwakuni, Japan, the Marine Corps felt it was important to send me to a Broadcast Management course on my way to AFN. That sounded wonderful at the time; however, I was not prepared for the humiliation that I was about to endure.

Remember, our first move from South Carolina to New Jersey? My wife and I were sitting on the bed wondering why God sent us here as tears poured down our faces.

This was our second move together as a family. We drove to Maryland, checked into a hotel and I was off to the schoolhouse for a few weeks of "fun." Imagine her thoughts as I put on my Dress "C" uniform, walked out to a wonderful August morning sky, and exclaimed what an amazing journey we were about to embark upon.

That night, I walked into the hotel room with all the joy ripped from my soul. Again, being the emotional guy in the family, tears poured out, "I can't do this…"

"What happened?" she asked, which has been a recurring question from my wife for the past 30 years.

The Unexpected Happened…Again!

Well, I walked into the huge DINFOS building expecting to bring all of my staff sergeant experience to the plate with fellow public affairs non-commissioned officers.

I made my way to the broadcast management room about 30 minutes early. There were several sailors and airmen in the dimly lit room holding conversations around tables put together in a horseshoe shape.

"I don't mean to interrupt, senior chief," I said assertively as everyone in the room stopped to look at me.

The Navy senior chief looked at me puzzled and asked, "Who are you?"

"Staff sergeant LT. I drove in from New Jersey and have orders to AFN Iwakuni," I shot back quickly.

"What detachments have you been with?" he asked, walking over to grab my orders.

"This will be my first assignment," I said in a softer tone, embarrassed that the others in the room would get to know my history.

"No wonder I've never heard of you," he said in a very concerned tone. He then turned to everyone in the room and said, "you see what the Marines do, they send these guys here with no experience thinking we can teach them how to run a detachment. It makes me sick."

He then turned and asked me when I graduated from DINFOS. That was the one question that haunted me for the rest of my career in Public Affairs. They didn't know that I had won several awards for my writing and graphics without the school. They only wanted to know my qualifications.

This seems to be the case in many aspects of life, doesn't it? I remember asking a Christian organization what I would need to become a missionary overseas. The answer was something like "get your four-year degree completed, then make sure you attend one of these seminaries. We will then have you go through a 1-year prep school followed by a few years of learning the ins and outs of pastoring along with raising support money." Phew, I thought. Too many years. Too much book knowledge.

After about a 30-minute interrogation process by the Navy senior chief, everyone in the room pretty much knew my full story. I was the outcast—AGAIN! "Why Lord?" I asked under my breath.

Of course, to start the day, everyone then proceeded to introduce themselves and their background in broadcasting. I remember hearing 10 years at Ramstein, 15 years in Wiesbaden, 12 years in Osan, and more. These folks knew their stuff. I was impressed.

"I have a Master's Degree, have served in detachments around the world and have won a Thomas Jefferson award for the best broadcaster in the Department of Defense," said the senior chief.

He then played his award-winning stories back-to-back, just to be sure that everyone knew how "amazing" our instructor was.

We were only about two hours into the first day when a very nice lady in her mid-40s walked into the room, "can I speak to staff sergeant LT, please?"

I jumped out of my seat and walked out of the room with her as quickly as possible. It was such a relief to be out of that room.

"Umm, staff sergeant, the senior chief asked me to look for your file that supposedly is with us," she said with a wonderful smile. "We don't have you as a graduate here."

"Ok, yes, I received the MOS with the official six-month on-the-job training program," I answered with confusion.

"Yes, well, you see, we need to have an English Diagnostic Test on file for our records," she said apologetically. "It is a simple test and will only take about 30 minutes."

"Now?" I asked.

She took me into an empty room, gave me the "bubble" test which I completed in about 10 minutes.

After grading it she said, "Staff sergeant LT, I have been here for nearly 20 years, and this was the best score I have ever seen on this test."

Deep inside, I knew that the senior chief was out to get me. Why would he report me to the front office after reviewing my record? Why put down a Marine in front of all my peers?

It was a very painful few weeks. Each assignment brought me to my knees. Most of the terminology and issues discussed forced me to remain very quiet, which is difficult for a guy like me who loves to talk and engage people.

Our final project was the killer for my career, in my opinion. We had to take all we learned about our future detachment and provide a brief about it to the commanding officer of the school. He was a full-bird colonel in the Air Force who everyone said looked and acted like Mr. Spock from the original *Star Trek* series.

Those with years of experience shared how difficult he was. With all of their time in broadcasting, one would think they could breeze through this brief with no issues at all. Yet, they expressed fear and anxiety leading up to that final day. The reality was, if you couldn't provide a comprehensive explanation of your detachment, your orders would be canceled.

The day of our briefing arrived. My wife and I prayed as if our very lives depended on it. I had no idea what I was doing.

I spent late night hours with very little sleep putting together a presentation that made no sense to me whatsoever.

When it came time for me to present, I gave a brief introduction, and then I clicked through my brief and talked faster than an auctioneer trying to sell cattle.

"Staff sergeant, you can stop right there," the colonel said with a concerned look on his face. "You don't know what you are talking about, do you?"

I knew this was the moment he would let me know that my orders would change.

"Sir, I do not, but I promise you that when I get to AFN Iwakuni, I will pour out every ounce of my being into not only learning how to operate one, but will make it award-winning."

I looked at the senior chief who appeared to smile as if he sensed my demise was forthcoming.

The colonel whispered to the board of directors and looked at me and said "I know you will staff sergeant. That is the best (bleeping) power point presentation I have ever seen in my life. Could you please stop by after we finish and teach me how in the world you put this together. Absolutely brilliant!"

Wow, not what I expected! I will never, ever forget those words or the promise I made.

You see, I figured I might as well go down swinging. So, in

preparation for my presentation, I went to the library that week and grabbed a page-by-page tutorial on how to create amazing powerpoint presentations.

No one really did this. They would just put some text in, drop a photo here and there and talk. Not me.

Each pie chart had a color. Whenever I would show a change in statistics, the colors on the chart would change and I would have arrows flying in to point to each aspect highlighted. It was fun to put together, and another building block for the future of *And We Know*.

My fear turned to joy as my wife and two small boys attended the small graduation ceremony. We flew to Iwakuni anticipating that it would be one of the greatest experiences in our lives.

AFN Iwakuni

"Good morning, everyone, I'm Staff Sgt. LT and am delighted to help you; however, for the next two weeks, I will be Private LT."

There was one soldier, several Marines and two sailors.

You should have seen their faces at that announcement. I figured the only way to learn broadcasting was to see how it felt on the front lines. How in the world can you be an effective leader, when you have no clue how the folks you lead feel each day.

They ran me into the dirt. I loved it. They pushed me hard on each story, confirmed deadlines were met and so much more. We were required to have a 10-minute newscast done each day by 1700, which would air at 1800.

That was only one aspect. We also had a live radio show in the morning and afternoon (more prep for *And We Know*). I would run the show at night for two hours to get the hang of it. It was the "Oldies but Goodies" show. Each night, I would open with Barry White's *Love Theme*, and say "Good evening Iwakuni, this is Staff Sgt. LT, coming at you live with 40,000 watts of power. Just sit back, relax and enjoy your favorite hits from the '70s and '80s. We will have special guests with us tonight including Wolf Man Jack and Elvis Presley."

I would talk like Wolfman Jack and Elvis during the show as

if they were in the room with me. It was fun. What wasn't fun, though, was constantly opening the radio room and screaming "Help, I forgot which button to push to go to commercials... someone help me!" My amazing sailors were always around waiting to assist during my training.

After two weeks of "broadcast bootcamp" I took over as the operations supervisor not knowing the senior chief would depart a few months later for Officer's Candidate School.

Marine Master Sgt. Parker arrived as his replacement, and he was one of the toughest leaders we ever came across. If our news segments weren't perfect, he would make us reshoot them. To be honest, none of us liked him.

I was bumped down to News Chief as we no longer needed an Ops Supervisor. We had a crew of three putting together the nightly news. Our combined experience in broadcasting was 18 months.

On one particular day, we covered the new trash receptacles on base and went on a helicopter ride with one of the squadrons. We actually reshot the trash scenes to make it livelier. This, combined with talking to the pilot through our microphone device while he was flying, made us feel that it could be a perfect opportunity for an award submission.

About one month after we sent in our tapes for the Thomas Jefferson awards programs, we received word that our newscast was #1 in the Marine Corps and that it would go up against everyone in the Department of Defense—including those large detachments with my peers who had master's degrees in Journalism and Broadcasting.

On a Sunday morning, a few weeks later, I was sitting in church with my family listening to Pastor Zane Abbott preach. Remember him?

The back of the church opened, and Master Sgt. Parker walked in, made his way to my pew and sat down next to me. Now, he was a Christian also, so I thought "Praise the Lord, he is attending our church now."

He leaned over during the sermon and said, "Staff sergeant,

guess what?" he asked with a smile. "I just got the message. Your crew has placed first in the Broadcast News submission for the Department of Defense."

I could only imagine how the Senior Chief at DINFOS felt when he saw my name as a winner.

Immediately following this victory, I was promoted to gunnery sergeant and transferred to the Iwakuni Public Affairs Office to become the chief.

Not only was I put in charge of the weekly news magazine, but they also asked if I would be able to create a website for the base.

"We heard you make things happen, gunny," said my new boss, a Marine Major. "Give the IT department a call and see if you can figure it out."

Again, I found a book on how to create websites and not only had one up and running, but as folks entered the site, they were greeted by a pilot staring out as if in a *Top Gun* movie, and then they would hear the pilot request to land. The air traffic control tower would tell you that you were cleared to land. And then an F-18 would shoot across the screen unveiling the website.

One year later, we won the award for the best website in the U.S. Marine Corps.

Blessed is the man who walks not in the counsel of the ungodly,
Nor stands in the path of sinners,
Nor sits in the seat of the scornful;
But his delight is in the law of the LORD,
And in His law he meditates day and night.
He shall be like a tree planted by the rivers of water,
That brings forth its fruit in its season,
Whose leaf also shall not wither;
And whatever he does shall prosper. —Psalm 1:1-3

Who knew that years later, our Heavenly Father would use all these experiences in my life to launch a popular broadcast program called *And We Know*.

Each news brief requires a graphic thumbnail, a presentation with the news for the day or week, a broadcast voice for radio podcasting, a website to post the information and the most important ingredient—trusting the LORD to have it prepared properly for all to see.

He can guide your life in a similar way. I have often heard people speak of His goodness and greatness and how He has worked amazing things throughout many lives. Yet, there seem to be many who are puzzled and say, "Why doesn't this happen to me?"

When folks ask how they can get their lives together, I simply tell them, "Dedication and hard work! And surrender to the Lord who can make it happen." Notice it takes dedication and long hours and a loving touch from the Holy Spirit.

The hand of the diligent will rule,
But the lazy man will be put to forced labor.—Proverbs 12:24

Chapter Eleven

Ain't God Good?

But thanks be to God, which giveth us the victory through our Lord Jesus Christ. Therefore, my beloved brethren, be ye steadfast, unmovable, always abounding in the work of the Lord, forasmuch as ye know that your labor is not in vain in the Lord.

[1 Corinthians 15: 57-58]

At the same time that the Lord was preparing me in Iwakuni, Japan, for the life of broadcasting "Truth" to the world, He was also calling me to a deeper love of His Word.

My wife and I were still newlyweds (first seven years of marriage), had two young sons, had traveled twice together around the world, and I had already learned several new jobs in a very short time span. Added to all of this was my constant learning from the Bible teachers on BBN Radio and other channels.

Up to this point, we've experienced the "Walking Marine," the Marine Corps Ball at the Taj Mahal, the storm that was "cleared" for our Garden State Games, and so much more. The one thing we lacked in this new life together was true Christian fellowship.

The way the Lord worked out "all things for good" to get us in the doors of Faith Baptist Church in Iwakuni was absolutely amazing.

"You really shoot that pistol well man," said Staff Sgt. Webster, as we completed the rapid-fire portion of our pistol qualification.

I thought, "wait a minute, how did he know I shot so well during rapid-fire when he is supposed to be watching *his* target?"

The young black Marine fired again and following that round he said, "brother, you need to show me how you do that."

As I turned in disbelief, he was looking at me with a smile that made me simply think, "I want to be happy, like that guy."

We continued with some small talk about where we were from, where we worked, how long we had been in the Corps and where we were on the gunny list.

As we walked out of the pistol range area, he headed over to his van which had huge text, "Faith Baptist Church," on the side. It was at that very moment I knew why he was so kind and had a special "magnetism" around everyone he met. He expressed a joy that could only come from a sincere love for our Heavenly Father and for people and was not ashamed of it.

For whosoever shall be of me and of my words, of him shall the Son of man be, when he shall come in his own glory, and in his Father's, and of the holy angels.—Luke 9:26

I was hooked. This guy was the real deal. At that moment, I was reminded that I needed to get my family plugged into a local assembly and had been praying for this very thing—to find a faithful, vibrant, Bible-believing church.

Well, it didn't take long for this prayer to be answered. My new friend from the range, Staff Sgt. Webster, invited us to Sunday worship and was looking forward to our visit.

We will never forget walking into this very small church which had about ten long pews, a tiny piano, and many families all dressed up and filled with an inexpressible joy.

You can imagine how we felt walking into this place filled with strangers, yet the people were so friendly and as each family introduced themselves, we felt that kindred connection that genuine believers often feel when meeting for the first time—like we had been friends for years.

These same families recalled meeting us for the first time and noticing that my arms and face were orange[4]. That's right, I finished a bodybuilding competition the night before at the base

---

4 It is a common practice for bodybuilders to use a form of spray-on tanning, in order to help accentuate their muscle definition in competitions. But this tanning spray would often leave the skin with a noticable orange tone.

theater where I placed first in the lightweight division. Posing in small briefs on stage one day and dressed for Jesus the next. That's a picture for ya!

Well, to my surprise, they started by requesting we open our hymnals to sing. Oh, how I love the old hymns. There is something to be said about the soul-stirring words of faith and hope, trial and triumph, and Christ and the Cross that pour out of these hymns, and the amazing stories of their saintly authors behind almost every hymn written. One of our favorites is "Trust and Obey."

They remind me of the many times I would take the purple bus to the local church in the 70s. We would learn about Jesus from the Sunday School teacher, color the books that matched what we had learned—like Joseph's Coat of Many Colors. After Sunday School, we would make our way to the large sanctuary, and they would begin worship service with a full choir and hymns.

Following the music, the "Smiling Marine", Staff Sgt., Jimmy Webster, made the announcements. Jimmy appeared as if he had walked up with Moses on the mountain and came down to share his love for God. He glowed.

After the announcements, they took the offering, we prayed and everyone took out their Bibles, pulled out notebooks, clicked their pens and sat up straight, anticipating what was about to unfold—the *sermon*.

A very short, gray-haired man in his 60s walked up to the pulpit. He had on a suit that appeared to be worn by a Forbes 500 CEO. He looked as if he was prepared for a final uniform inspection in the Corps.

As he looked at the congregation, he smiled, put his glasses on, opened his Bible and we entered what I would describe as "Heaven on Earth."

"Well Glory!" he said while opening his Bible. "Turn in your Bibles to Matthew, chapter 1."

As he expounded on the Bible, I found myself in awe.

He read each verse and broke it down in the simplest fashion. He used an outline for his sermon to help us remember his key points. And he preached verse-by-verse, expounding the Scriptures, and making relevant application to his hearers. This was something foreign to me. It is called expository preaching.

Every now and then, he would pull the glasses off his face, fold them in his hand, turn to the right or the left and start preaching with authority, bringing one foot forward with fists clasped, leaning out into the air as if he was reaching for heaven.

My heart was burning. Each minute that went by was filled with awe and wonder. Tears wouldn't stop flowing.

As soon as he completed the sermon, he made his way to the back to warmly greet folks as they left the church. Somehow, we made it to the car without him seeing us. I didn't want him to see my orange hands.

We were parked across the street in the visitor's parking lot. As I looked back to the church, Pastor Abbott's wife Burma, tapped him on the shoulder and pointed at us and must have said, "Go talk to that new family."

The next thing I knew we were walking across the street towards him. He shook my hand and asked where we came from, all the while looking at my hands. To this day I think he was telling himself, "Good Lord, who did You send me this time? This guy is orange."

Throughout that year Pastor Abbott worked diligently and masterfully to influence the spiritual growth of each young Marine the Lord sent his way. He would start by asking us to pray in public in front of everyone on Sunday night before going home. He would invite us to men's Bible study on Thursday nights to watch videos about Israel. After watching and praying, he would eventually ask us to pray with him before Sunday night service in a small, tiny building behind the main sanctuary. That carpeted, small room had one small lamp on an end table and three small chairs.

We would discuss the current hardships in the church, talk about the sermon for that night and three or four of us would get on our knees and go around the room, each one of us praying.

About 12 months later, we were having dinner and going over

the financials, as I was the new treasurer.

"Son, Jimmy was telling me that you have been chomping at the bit to do more preaching," he said, changing his demeanor from joking with our sons to a serious tone with me.

"Yes sir," I said. "I have been hoping to preach someday," I added with a little hesitation.

Pastor Abbott pulled out his schedule book. He opened it and said, "how about this Sunday night?"

It was Tuesday. I said "Okay."

Sermon Preparation
That night, I pulled out commentaries borrowed from the pastor's office, got my Strong's Concordance out and picked a verse that I had been thinking about for a long time.

And said, Verily I say unto you, except ye be converted, and become as little children, ye shall not enter into the kingdom of heaven.—Matthew 18:3

At this moment, without looking it up, the Greek word "epistrepho" still lingers in my mind. It translates "converted" in that verse and it means to completely turn around and go the other direction. Epistrepho was only used once.

That Sunday night, my mind was calm, the notes were ready, but my body was shaking in fear. During the sermon, I could sense the Holy Spirit guiding my every word. What brought me added comfort and encouragement was seeing Pastor Abbott smiling and saying "Amen" at key points in my sermon. I realized that standing at the pulpit, teaching the Word of God was where I wanted to stay for the rest of my life.

After the sermon, many men came up to give a quick side hug and praise the Lord for the sermon. Walking outside, I noticed my wife talking with Pastor Abbott and his wife, Burma, in a very excited tone of voice.

He asked her how long I had prepared for the sermon. She proceeded to tell him that I had books opened all over the living room and dining room and was constantly reading and writing each night after work.

As I walked up, he said, "Son, I have been doing this for many years and have a good sense of a man's gift for preaching."

"Most men who preach for the first time normally get a "D" or might make it to a "C+"; however, your sermon tonight was an "A-" because you missed one minor thing. You told everyone to go to Matthew 18:3, when you should have first sent them to Matthew 18, giving them time to find the chapter, then take them to the verse."

"You could be called to preach, but I don't want to get in front of the Lord," he added, in a tone of confidence.

Great wisdom, I thought. He saw something in me, but didn't want to give me a false hope that would lead me to prematurely leave the Marines and run to seminary.

That night, I walked into my bedroom and felt an enormous weight on my soul. My chest was filled with a pain that was not physical, but spiritual.

"Joe, something's wrong with me," I said on the phone to my good friend who preached once a month.

He asked me what I had done that night. He wanted a play-by-play after the sermon.

"When getting the car I asked my wife how she liked my preaching and…"

"Oh, I know what you did, LT," cutting me off. "Never say 'my sermon,'" he said with a chuckle. "That is pride my friend. The Lord used you as a vessel for *His* work."

After we prayed together, the pressure on my heart was removed. It was a great lesson in humility.

Every Friday night the pastor had seven "Preachers" over to continue his training. There was nothing like entering a house in Japan filled with Southern-style furniture and the smell of coffee filling the air.

Mrs. Abbott had a way of filling our bellies with her delicious baked goods while Pastor Abbott filled our souls with the training we all needed and yearned for. He would often play video tapes from Pastor Lee Robeson out of Tennessee Temple University.

We never wanted to leave their house. Many men would come up with excuses to stop by, just so we could be with the Abbott family.

On one special occasion, they had me over to discuss a few things about the treasury report. During a break, they popped in a video called "Precious Memories," connected with a name I had never heard of—Bill and Gloria Gaither.

I was so moved and they let me borrow a few of the Gaither videos so we could play them at home. That's when we discovered a group that would forever change our hearts through music—*The Isaacs*. Soon, we had our sons singing their songs and learning harmony by singing their song "Children, go where I send thee."

More than 20 years later, in 2022, we would end up in one of the Isaac's homes in Tennessee, singing and praying together, thanks to *And We Know*, but more so, thanks to the Abbotts.

Although my wife was convinced she was a believer in Christ, Pastor Abbott would spend time with her in our home and really discipled her to understand just who Jesus really is. He ended up leading her in prayer to receive Christ and baptized her in the Pacific Ocean. The time he invested in us was so amazing, she ended up calling him "My Papa."

When we received word that Pastor Abbott had left us to enter his eternal rest in March of 2021, we mourned for a long time.

Our Move to Virginia

After departing Iwakuni, we ended up in a small Southern Baptist Conservative church in Newport News, VA. We were immediately welcomed and felt like family.

The church was growing fast. We provided youth group support and learned about Evangelism Explosion (EE). It was such a great blessing to be part of this body of believers. The women were a close-knit group and spent much time together in prayer and the congregation was exemplary in how they demonstrated their love for one another as well as the lost.

On one Wednesday night, during our mid-week church service, I was sharing with my wife how embarrassed I was with the "strange bright pants with flared endings" she was wearing. They were called "capri" pants. I mumbled to her something along the lines that they didn't seem very "godly" to me.

Now before you ladies put this book away after reading that, hang in there…our Father has a great sense of humor.

She got up in the middle of the sermon, grabbed the keys and left. I started walking home right after the service. Dennis caught up with me and asked what was going on.

When I told him the story of the pants, he was laughing so hard and proceeded to give me a ride home. He must have told his wife.

On the next Sunday morning, we arrived about 15 minutes early; however, the church parking lot was already full and everyone seemed to be in church.

We thought this was strange. When I opened the doors to the church, about 15 women were standing in line sticking out their legs all wearing Capri pants.

They high-fived my wife as we walked in. I got mine. That was PAYBACK. But more than that, it was LOVE. It was GRACE.

The pastor provided several opportunities for me to preach on Sunday morning and evening services. He also asked if I could lead the men's weekly Bible studies each Saturday.

Those are some of my greatest times in ministry—spending time with men in the Word of God. I simply used the same model that we had in Iwakuni at Faith Baptist Church.

I would play the "Faith Lessons" by Ray Vander Laan which had a tremendous impact on my faith. In each lesson, Ray takes a group of people throughout Israel and Europe into a place with significant impact from the Bible. My favorite lesson from that entire series was Elijah and his encounter with the prophets of Baal—and how God showed up in a mighty way! (1 Kings 18).

We would discuss the lesson, open our Bibles to share what our Father in Heaven conveyed to us in our own lives which would lead us to prayer.

That church is also where I learned and taught Evangelism

Explosion. This model of teaching is tremendous and simply provided a perfect outline on how to share the gospel for those of us who wanted guidance.

Each week we would visit a family or person who stopped by to visit the church that Sunday. It was always a treat to take new Christians on their first visit to a home to share about Jesus. We had large 200-pound bodybuilders who were in deep fear over this concept. I would always remind them that God that orchestrates these "encounters" and the Holy Spirit would provide the most amazing experiences from these visits. One bodybuilder couldn't stop talking about one such home visit as he saw the Lord melt a couple's heart for Jesus.

The Greatest Visit of All

I sense that it is important to expound upon this portion of the chapter, providing one of the greatest moments in my evangelism experience. Although this book can't provide a play-by-play of every significant event in my life, these few should help define my love for seeing folks come to Christ.

During one particular Evangelism Explosion training visitation night, I had two men about 20 years older than me and one other younger man. We pulled into the parking lot of the family and, as always, prayed together for this visit to be directed by the Holy Spirit, before we went inside.

We discovered that many family members were in the house. They had parents, sisters, cousins and friends over. It was a full house.

Since I was providing the training, I asked the older men to take over, ask questions and look for a "springboard" to the evangelism part of our time with this family.

Well, after an hour of chit chat, the older men kept rubbing their hands nervously, sweat was pouring from their foreheads and I could see their stress levels were increasing.

Finally, Dennis said, "Well, it seems we have had a great time talking with you all. LT, do you have anything to say? You're kinda quiet over there."

Yup. You guessed it. He punted the ball to me.

"Well, you know it has been so wonderful talking with you all tonight," I said to the family with confidence. "As you know, we stopped by to visit your family as a thank you for visiting our church family on Sunday."

"The most important thing is we want to convey to those who visit our church is an understanding as to why we meet and Who it is we are talking about," I added. "So many churches seem to have a country club approach with no real purpose."

"Now, for some reason, I sense the Lord really wants me to talk to that young man in the middle of the living room, but I could be wrong," I said, looking right at Steve, a man I had never met in my life. He just happened to be visiting the parents of his girlfriend.

As those words flowed out of my mouth, the entire room filled up with tears. The wailing cry of his girlfriend's mom was quite telling, for sure. The women in the house ran to the closet and pulled out tissues, as they were all crying.

Only once in my life have I had the boldness to share so directly what was in my mind. This is not a common practice for me at all. Normally, our EE approach is we talk to everyone in the room, ask the two generic questions and wait for answers.

Steve could have said, "Sir, you can leave 'cause I don't want to hear it," but he didn't.

"Steve, do you mind if I ask you a couple of questions?"

He nodded with affirmation.

"If you die today, do you know for certain that you are going to heaven, or is that something you are still working on?" I asked softly as the women continued to cry, passing tissues as they sat next to him.

"Sir, I don't really know," he said in his soft Southern accent.

"Thank you for being honest," I said. "When you leave this world, you will see God. If He asks you why He should let you into His heaven, what would you say?"

"I don't know," Steve answered with confusion. The crying continued.

We then proceeded through the Gospel message.

At the end of the teaching, I asked, "Steve, the Bible says, 'for the wages of sin is death, but the gift of God is eternal life through Jesus Christ' (Romans 6:23)."

"It is a gift. When you received a gift for your birthday, do you reach into your pocket to pay for it? No, right? So, when God says eternal life is a gift, that is something you can freely accept."

"Would you like to receive the free gift of eternal life?" I asked. Steve said "yes" and the entire room continued in a fresh set of tears.

"Do you believe Jesus died on the cross and rose again after three days?" I asked.

"Yes," he said.

We prayed in the living room, and he was thankful.

What an amazing night. Before we left, Steve's girlfriend was glowing and her mom pulled me aside and said, "we have been praying for Steve and hoping he would ask my daughter to marry her."

It didn't end there. He was a newborn Christian. He needed to grow. He noticed that something strange was happening to his mind, and the following week at church Steve wanted to talk.

"My girlfriend has been acting strange lately," he said right after Wednesday night service. "When she woke up, she didn't say anything, got out of bed and walked out of the room. She never does that."

So, immediately I caught on to something very revealing in that statement. Did you, reader, happen to catch it?

"Steve, did you say she woke up next to you?" I asked. "So, she sleeps with you in the same bed?"

He looked at me puzzled and asked me if that was wrong and went on to explain that since he prayed to receive Christ, he felt awkward sleeping with her and didn't know how to explain it. This was also the reason she was acting so strange.

She thought he was going to break up with her because he felt a sense of guilt having intimacy without being married.

As Pastor Abbott would say to that, "Well Glory."

Well, we went straight to the Bible. I took him to the Epistle of First Corinthians.

Now concerning the things whereof ye wrote unto me: It is good for a man not to touch a woman. Nevertheless, to avoid fornication, let every man have his own wife, and let every woman have her own husband.—1 Corinthians 7:1-2

My friend, Jimmy from Iwakuni, happened to be visiting me that night as he was in Virginia for Marine Corps training. We had a blast together.

Steve was very receptive to what we explained to him from God's Word. Yet, he was puzzled about how her family, supposedly Christians, would allow him to live with their daughter if they knew about these Scripture verses prohibiting intimacy before marriage.

We were careful to explain that he had a decision to make now, and that it was best to sit down with his girlfriend, share what he he learned and make sure not to judge anyone.

My phone rang a few nights later. A call I will never forget.

"LT, I couldn't do it anymore," he said, sounding very depressed. "I let her know that it was not good for us to sleep together outside of marriage. She got angry and started crying. I packed my stuff and left."

We prayed together, and I asked if he had a desire for her. "The Bible says in 1 Corinthians 7:9, *But if they cannot contain, let them marry: for it is better to marry than to burn.*"

He said, "Of course I do! I love her and it is hard for me to keep my hands off her."

"Well, Steve, you have a decision to make then, don't you?" I asked him again, before we prayed.

Two days later, the girlfriend's mom was waiting for me at the door of the church when we pulled into the parking lot.

"Uh oh, I think I am in big trouble," letting my wife know that she looked angry. "She must be waiting for me. I bet they are angry because her daughter lost her boyfriend."

I got out of the minivan and sure enough, Steve's girlfriend's mom started making a beeline for me. As she got closer, I backed up and apologized. My plan was to get back in the van and go home as I told my wife to please wait so we could escape.

"LT, where are you going?" she asked me with a huge smile. Opening her arms to hug me she said, "I don't know how you did it, but we are so happy right now."

"What do you mean?" I asked with a puzzled look on my face. "You're not mad at me?"

"What are you talking about LT?" she said with a deep Southern drawl. "Didn't you hear? Steve asked my daughter to marry him yesterday. They are getting married next week. Praise the Lord!"

Ain't God good? It is amazing how applying the Word of God into our lives can be so effective and so liberating! The world's system is completely opposite to what we hold to be true. This has been a staple for my growth in my ministry and impact throughout my life.

Throughout the years, our Heavenly Father has guided me to spend quality time with men who are willing to study the Word of God, with a geniuine desire to learn and grow, and apply it daily in their lives. The Lord slowly took me through all types of training in many areas including leading a pastor search committee, taking over the pastoral duties after a pastor failed in a church, leading worship in several churches, preaching for chaplains on military bases, and so much more.

Reaching the World
This part of my story would not be complete without conveying one final area of life that only our gracious Heavenly Father could orchestrate.

After all of the years of training in ministry, God had one huge lesson for me to learn which prepared me for *And We Know* to reach people throughout the entire world.

He sent me to Hong Kong for a worldwide youth conference in 2006. We flew there with our youth group when I was stationed on Okinawa as a Master Sergeant in the Corps.

At the time, Hong Kong was beautiful and was not as corrupt as it is today. We enjoyed the markets, the food and talking to many folks who appeared Asian but spoke with a wonderful British accent.

Our event was held in the Shaw Auditorium where more than 5,000 youth and leaders flew in from around the world. We met so many wonderful people from Jamaica, Thailand, the Philippines, different countries in Africa, Germany, and many more.

After listening to wonderful worship from Jamaica, we were told to head to smaller groups in rooms already assigned for us. These groups would be around 300 people each.

After a small lesson, the leader of the room, a man from Morocco, asked for ten volunteers to lead a group of 30 to discuss what was learned during the lesson.

Of course, Marines love to lead, so I volunteered. The lesson for the day was wrapped around sheep, so we spent 30 minutes talking about sheep.

Things didn't go as planned though. My group kept looking over their shoulder as they noticed other groups were laughing, doing skits and just "having fun."

As a matter of fact, three young ladies, stood up and left for another group because I was boring.

The following day I devised a plan not to go back into that room due to my failure as a teacher. After worship, there was a great hiding place to simply sit and read. No one would find me.

Well, during worship, a young man from India sat next to me, pulled out a photo album and shared pictures of his congregation and told me stories that brought tears to my eyes.

As he continued to share the photos, we walked together down the hallway where everyone was gathering for small groups.

Realizing we were near our room, I panicked, thanked the man for his testimony, shook his hand, and turned to run away.

"LT, LT!" was heard as loud as thunder from a thick Moroccan accent. I knew who it was, since we had talked a day earlier about my life as a Christian in the Marines.

I turned to acknowledge him with, "Yes sir, do you need something?" still planning my escape.

"LT, your story touched me so much yesterday."

You know these people are from all over the world, and I want them to hear your story about living for Christ and being a Marine at the same time," he said with passion.

"Can you please take 15 minutes to share this with everyone before we break up into smaller groups?" he asked with a hint of grace that no one can say "no" to.

So much for my escape plan.

About five minutes later everyone was seated, and I was jotting down any thoughts that crossed my mind while turning in my Bible to find an appropriate verse. And then it hit me… "Tell the story about the gunny."

The young man from Morocco painted a picture of me as if he had met the likes of Dwight L. Moody. He had me walk up and stand facing the assembly, and the first thing I noticed were the faces I disappointed the day before. But, the Lord gave me the strength to share my story.

"Ten years ago, I made a decision to live for Christ each day by placing a Bible on my desk," I said while visibly shaking, looking at an audience filled with every race on the earth. "That decision was not as easy as one might expect."

"The first week I did this, a gunnery sergeant walked into my office and said, 'Sergeant, you are not allowed to have a Bible in your office!'"

I proceeded to let everyone know that I stood my ground and with respect told the gunny that under the Constitution of the United States, it was legal to have a Bible in my office.

He didn't like my response, and turning blood red, he screamed at me, "Follow me now, sergeant. We are going to see the sergeant major."

I followed him down a long hallway straight to an empty office. He slammed the door, got in my face and yelled at the top of his lungs.

"Who do you think you are, sergeant?" He continued to yell at me. Yet, through it all I felt peace.

There seemed to be a heavenly force field around me. I felt the Lord's divine protection. I recalled the biblical truth that it is a great thing to be persecuted in the name of Jesus.

It must have had an effect on the gunny as I heard, "what are you smiling about sergeant?" he asked, breathing hard from exhaustion.

"Well since you asked, gunny, I have to tell the truth," I said to prepare him for what was about to come out of my mouth.

"I was just thinking how wonderful it is that you are yelling at me, because the Bible expresses the comforting truth that there will be rewards in heaven for those who endure and persevere through the persecution that comes from those who hate Jesus."

"Get out of my face sergeant," he yelled.

Blessed are ye, when men shall revile you, and persecute you, and shall say all manner of evil against you falsely, for my sake. Rejoice, and be exceeding glad: for great is your reward in heaven: for so persecuted they the prophets which were before you.—Matthew 5: 11-12

Later that day, about 4 p.m., that gunny walked into my office and, in a gentler tone, asked me to come to his office.

"Sergeant, I apologize for the way I acted today. I was wrong," he said while looking into my eyes.

The next moment, I asked him a serious question that surprised even me.

"Gunny, why do you hate God?"

With tears beginning to well up in his eyes, he said, "The church kicked me out because my wife and I got a divorce."

For the next 30 minutes, I talked to the gunny about God's grace and mercy. We talked much about salvation through Christ and I invited him to church.

I then proceeded to let the group in Hong Kong know that Jesus, in Matthew chapter 8, marveled at the faith of the centurion, a man in the military. As a matter of fact, Jesus says that He had seen no greater faith. Did Jesus ask him to stop wearing the uniform? No.

Immediately following this presentation, we prayed and broke into our small groups. The same group who was bored the day before, ran to me in tears. One after another, folks from India, the Philippines, Thailand, Hong Kong, and many other countries wanted to let me know the same thing.

"Brother LT, I was always scared to have my Bible at work, but after today, I will put the Bible on my desk."

When Jesus heard it, he marveled, and said to them that followed, Verily I say unto you, I have not found so great faith, no, not in Israel. —Matthew 8:10

CHAPTER TWELVE

Five Kids, No Future Job...Walk On Good Water

And he said, Come. And when Peter was come down out of the ship, he walked on the water, to go to Jesus.
[Matthew 14:29]

Every member of the human race has experienced, or will encounter, trials and suffering in their lives. The question is, what will they do during those times? How will they respond? Where will they turn for wisdom, guidance, healing? Many have lost their job, or come down with a sickness that destroys their dreams, or simply live each day wondering if there is more to life than "this life I am leading that seems to have little meaning or purpose."

I am often looked to for answers to their burning question; "What do I need to do to fix my current difficult life situation?" That can be a challenge for many to answer, yet I have found the simplest answer is a question: "Are you living each day seeking our Heavenly Father through His Son, Jesus Christ, as revealed in the Bible."

What I mean here is not an intellectual knowledge or understanding, but this is about an intimate relationship with the Son of God. I've shared with folks, many times, that while there have been times that I have struggled in my Christian journey, it seems that there hasn't been a day go by that I don't think of Jesus and talk to Him, even when I have been very upset or disappointed with my situation at the time.

One thing I do know, though, whenever He is ready for me to make a big move, our Heavenly Father normally gives me a huge distaste for the world's attractions that can so easily plague and distract my daily life. It is like a tooth that hurts slightly, but if you don't deal with it in a timely manner, your pain increases and you end up in a dental crisis.

Making a Huge Life Change
The Marine Corps was starting to become a very difficult life for me the final few years that led to my 20-year retirement. I would often ask the Lord, "Could you please give me a "powerpoint presentation" unveiling my future? I am ready to get out, but not sure what to do?"

What I've found enlightening, in God's answer to my prayers, is that a few things occurred that led me to consider doing something new. This didn't make much sense at the time, but it all came together at the time I was able to let my wife know in 2008, "I think it is time for me to retire from the Corps." I expected her to say no, but she simply replied, "Ok."

That announcement was made right after the housing crash and right before "O" became the 44th president of the United States. We also had five kids at home in a house we purchased one month before the crash, with one girl who was only eight months old.

The Dream That Set the Stage
In 2007, two years prior to my retirement, my Marine major looked at me and said "Master sergeant, I need you to fill a billet in Germany for a joint military exercise. They need a Public Affairs chief."

"Now, ma'am, there is no reason for me to go to Europe," I answered. "Can't they fill it with someone else?"

This happened right after we discovered that our final child was on the way, so it was the wrong time for me to leave the family. The major said, "You're going. They need you."

She was bragging to the higher ups about my work as a chief for the tsunami in Thailand and more, so it was too late to change plans.

Against my will, I had been ordered to Germany. As soon as I arrived in Frankfurt, I was bused to base where I ended up in a small barracks room.

On one particular weekend, the U. S. Military took the Americans on a field trip to Heidelberg. After a short tour of the amazing Heidelberg castle, I sensed a need to pray. It was overwhelming. I happened to find a trail behind the castle that took me about 200 yards to the East of the castle on the mountain.

I soaked in the incredible scene around me while sitting on a large rock that sunny day. I noticed a beautiful river and a historic cobblestone street filled with people, and expressed to the Lord, "I would love to bring my family here someday."

About 12 months after this journey to Germany, I had a dream one night, where I found myself on the corner of a cobblestone street that reminded me of the one I saw in Heidelberg.

In my dream, I saw two giant men appear wearing soccer uniforms.

"We've been looking for you," one of them told me with a strong British accent. "We need a head coach for our team."

For some reason, I knew that they were with Manchester United, a team name that was well known from that area.

"When do you need me?" I asked, super excited about the offer.

"We will finish the paperwork in January, and have you start in February," the other player answered.

"Oh man, I am planning to be reenlisting soon and remain in the Marines," I said, disappointed as the words came out of my mouth.

They looked at each other and one of them said, "Oh well, if you change your mind give us a call," while handing me a rolled-up scroll about six inches long.

As soon as they handed the scroll to me, I was transported immediately into a moving train.

"Tickets please," said the conductor, as I noticed his fine 3-piece suit, antique-looking glasses, full grey beard and a hat that seemed to be from a bygone era.

I handed him my scroll. He took it, and as he unrolled it, it revealed a shining glow so bright it caught the attraction of everyone in the train car.

"Do you know what this is?" the conductor asked me, as if he was looking at a million-dollar lottery ticket. "Did you take this offer?"

"No sir, I am in the Marine Corps," I answered slowly.

"This is probably one of the greatest offers I have ever seen," he said, and everyone started clapping.

Then, in an instant, I found myself immediately back on the cobblestone street standing next to a distinguished-looking man wearing a robe who appeared to be one of the Apostles of Jesus. We were looking over a wall that had about a 15-foot drop off to the street below.

The next thing we saw was my family in our minivan driving to the edge of the wall, waving and smiling while saying, "Hi dad, this is great."

The van jumped the railing and was heading through the air down towards the street below. Emptiness filled my soul as all of this was happening in slow motion. The man next to me reached down, grabbed the van and brought it over the railing. They were safe. I woke up sweating and in tears. My dream felt so real!

Now, lesson learned here. Be careful when asking God for life-revealing "powerpoint presentations". You might just get an answer in a full movie production. Phew.

It didn't make sense at the time; however, it would become crystal clear over time.

Each day following those dreams became increasingly more painful. Our building caught on fire at the headquarters in Cherry Point, losing all of my awards and news articles collected throughout the years. We also had a Marine who complained of religious harassment in the office because the staff sergeant in the office who claimed to be a Christian counseled him that morning using the words, "We are in the light here in this office, and you are in the dark."

Since I was in charge of that office at the time and had a Bible on my desk, they threw me in the mix as part of the complaint and conducted a full investigation.

After two weeks of questioning everyone in the office, we received a call to report to the commanding officer. The staff sergeant and I walked in, had a seat and waited nervously for our "sentencing."

I say that rather lightly because whenever an investigation is launched on you in the military, your entire career seems like it's going to end. No matter how innocent you may be, this is not a position anyone wants to be in. I was close to 20 years in the Corps and didn't need this, especially as I was considering retirement.

"Well, we have the results of the investigation," the CO said looking at the two of us and the sergeant major who was sitting next to him. "I must say that this was all very eye opening for both of us."

Turning his gaze directly at me, the CO added, "Master sergeant, everyone in your office likes what you are doing over there. I have never seen anything like it. Whatever you are doing, keep it up," he ended, with a grin on his face.

Pausing, he looked at my staff sergeant and said, "Here is a letter from me to you staff sergeant," leaning over to hand him a one-page document with a few paragraphs on it. "This is not saved on my computer, nor will it go on your record."

The staff sergeant looked puzzled as he started reading the letter and said, "I have a problem with this, sir." I immediately requested to be dismissed so they could talk it over, privately.

Without going into too much detail, I will only say that I tried desperately to encourage the staff sergeant to order the removal of all of the posters and bumper stickers on his wall that was a visible expression of his moral stance on many issues. He felt the need to let everyone come to the understanding that if you haven't put your faith and trust in Jesus, you are going to hell. Oh yeah, that was all over his desk.

His answer to me was simple, but insightful. "Do you see the chapel on the base? It has a cross on it. If they can have a cross on this base, then I can have all of this in my workspace."

Again, it will take another book to tell you what happened to each person that acted this way; however, the drama only continues there.

The Final Straw
Remember, I mentioned earlier in this chapter that our Heavenly Father knows how to prepare us for our next assignment. Well, he allowed this next battle to end my willingness to reenlist in the Corps.

"Master sergeant, the commanding officer wants to see you," said the civilian secretary.

"Did he say why?" I asked, since this was not ordinary.

"Have a seat, master sergeant," said the sergeant major, who was in the CO's office waiting for my arrival. "It appears you have a lot of leadership issues over there."

That's how it got started. Instead of telling me what was going on, the two leaders started by accusing me of making poor decisions in the office.

"Your 19-year-old lance corporal stopped by to file a complaint stating that you are getting into his personal business."

"Is that right?" I answered with authority, realizing that I had just passed 20 years in the Corps. Now, I wasn't being disrespectful, but when you get past that mark, your ability to have conversations, Marine to Marine, change. If you understand this, say "Oorah!"

So, imagine me sitting there in that room, looking around in slow motion, realizing that this would be my last stand as a leader in the Corps. It was like a scene out of the movie A Few Good Men, when the lawyer (Tom Cruise) realized he is about to go on the attack of the Colonel (Jack Nicholson), knowing it was a "do or die" moment for him.

Think. Breathe. Get your mind in order. "Jesus, help me."

"Sir, did the young Marine tell you that he is married to his high school sweetheart who is pregnant and that the bank is about to repossess their car for not making the $400-a-month payments?" I asked them both.

He looked at the sergeant major and turned to me and said

"No, he did not."

"Did he tell you that I canceled my camping trip with my family to take him and his family to a car dealer to help them get a trade-in to knock down their payments?" I continued.

"Did he tell you that they were receiving notices to get their housing cleaned up or they would be removed?"

"Did he tell you that he was paying someone to cut his yard and couldn't afford it due to his bills and broken arm?"

"Did you know his lawn needed mowing, so we took care of it for him?"

After all of questions, the answer was classic. "Well, you could have asked the Staff sergeant to look into it all."

"You mean the one that you just had a problem with, who was calling several Marines to discuss their spiritual journey in life?"

I then said, "You both have just made it easy for me to finally move on. I am dropping off my retirement paperwork today. I appreciate you both on helping me with this," I added, then requested to be dismissed.

Later that day, the paperwork was done and submitted.

I was given a retirement date of January 31, 2009. That was my birthday. Pretty cool, right?

So, not long after this, the senior enlisted public affairs chief, responsible for our next set of orders gave me a call. He was a good friend of mine.

"Hey, master sergeant. I just got the word you are punching out," he said with a confused tone of voice.

"Yes, master gunnery sergeant. I have had enough. It's time to go," I answered without hesitation.

"Do you realize that you will be up for master gunnery sergeant next year (E-9)? I also had orders for you to head to Camp Pendleton and head to Afghanistan for 18 months."

Had I stayed in the Corps and followed through with that order, it would have been devastating to my family. This was such a crucial time in the lives of my two oldest sons and the prospect of my wife having to care for all of our five children without me for 18 months was not an option.

The "Superman Christian"

Of course, one can only imagine the craziness of moving out of the Marine Corps with five children and no future job in sight.

Many of my friends found out about my "ridiculous" decision and called to give me advice. Each phone call ended the same way.

"Wow, LT, you've got me thinking about this too," said a fellow master sergeant based in Virginia. "I don't have the faith that you have, but you make a lot of sense. I am tired, too."

The following year, five of my fellow staff, non-commissioned officers, retired and ended up finding amazing jobs in government service.

Finding a job, turned out to be a nightmare. In those days, I applied to numerous jobs through several government websites which were specific to each branch of the military. Today, you can apply through USAJOBS or other similar online portals and find availability in your field of work in any branch or on any military base.

Of course, my experience and comfort level was in Public Affairs in the Marines, but I was willing to expand my applications to all Navy and Army public affairs positions. After each application, I would contact the hiring officer to introduce myself and let them know that they could call anytime. One of my friends thought I was a shoo-in for a job in Yokosuka, Japan.

But that fizzled.

I then received a call from a Navy base in Washington related to a position where I would create videos for submarine commands. This seemed like a perfect job for me. A few weeks later, the hiring officer broke protocol and called my personal number while we were having breakfast.

"LT, I have to tell you that you are the perfect fit for us, but someone else, a lot younger than you outscored you in a few areas," he said with a sense of urgency.

He wanted me to know that it was difficult to get into a government service job as a GS-11 or GS-12 right out of retirement, but that it would be better for me to look for more leadership level jobs instead of a "worker bee."

During this process, our church friends would often inquire how we were doing. In the beginning, it was all about "God's got this." Each answer seemed filled with an overwhelming faith and confidence that our Heavenly Father was in control and guiding us in this process.

Now, as the days blended to weeks and weeks to months, my superman Christian shirt was fading. I got in the habit of wearing a shirt over it, and the desire to tell everyone that Jesus would help me find a job started to fade.

Yes, doubt started setting in. I even did something that was awkward for a Marine. I applied for an Army public affairs job in Mannheim, Germany. Desperation was really setting in. Now, I am not saying that the Army is a bad branch. Please don't misunderstand me. We, as Marines, tend to be more comfortable around our own.

Of course, I made the standard call to the representative there and gave him all my information. He was very quiet and I sensed from the call that he would never call back.

In December of 2008, one month prior to retirement, the sergeant major called me to his office.

"How are you, master sergeant?" he asked as he saw I appeared tired and depressed.

"Hangin' in there," I answered in a week tone of voice.

"Well, you know that you can take this paperwork to the major general at Camp Lejeune and have your retirement package pulled?" he said while handing the paperwork to me.

I took the paperwork and walked out thinking that maybe our Heavenly Father didn't want me to retire. I drove home to get some lunch to prepare for the 45-minute drive to Lejeune. It felt like the weight of the world was on my shoulders.

The paperwork I held felt like "Kryptonite" in my hands. It made me feel weak and I knew it was a spiritual heaviness. The Holy Spirit was tugging at me.

"OK, Father in Heaven," I said, sitting in my home office chair. "I won't do this. We will move to a small mobile home down the road, and I will apply for a job at the local shopping store."

As I said the words, "in Jesus' name, Amen" the phone rang.

"Master sergeant, someone called from Germany about a job interview," said my corporal from the office.

"Really! Give me the phone number," I yelled into the phone with an air of excitement in my voice.

Dialing that number was difficult. My fingers were shaking. A woman answered the phone. After introducing myself she said something I will never forget.

"Wow, that's the quickest return call I have ever received," Jane said.

"If you knew what I've been through for the past four months you would better understand my sense of urgency," I said without hesitation.

Interesting right? Most phone calls to a future employer are somewhat formal in nature. I let it all out.

She wanted to know what I had been through, so in true LT manner, she got the full story.

After some good back and forth discussion, as we were both settling into the conversation, I said, "I thought you already found someone for this position," referring to the rejection email they sent earlier explaining that they appreciated my application, though I didn't make the top of the list.

She said that the first five candidates did not work out and because I had called repeatedly, they decided to put me in the running along with the next top applicants.

We talked for over an hour. It was amazing as we covered many related subjects from how government service works, to life in the Army in Germany, to my entire family profile and more.

That day, I ran to the library, checked out a DVD on Germany, and we watched it together as a family.

My wife thought I was overdoing it since the interview was the next day and nothing about the position was guaranteed.

"I know this [job] is it," I said to a somewhat doubtful wife who knew me very well.

The following day they called me for the interview. They really drove home the point that this was a magazine editor position and that I was clearly overqualified.

To settle them down, I told them the following, "Each time a new second lieutenant checked in and worked alongside me as the new chief over the office, I would tell them if they wanted to make an impact as a new leader in the office to not bark out orders but instead seek to become a servant to each one of them. Provide the tools they need to make it…lead by example."

"You better serve a hot meal every day," would be my typical ending to such a story every time.

After completing this explanation, Jane said, "oh wow LT, what an amazing story. I love it."

We completed the interview and she knew that I would need an answer on the job soon, but reminded me that it would take several weeks before a decision would be made.

About four days later, after some lengthy prayer, my wife and I were driving around getting ready to make a move out of our comfortable life. We discussed how it would all work out if Germany didn't actually come to fruition.

Then, my phone rang.

"LT, this is Jane," she sounded happy and cautious. "I am not supposed to do this because anything can happen in the HR department, but I know that you need answers soon in light of your upcoming retirement."

"You got the job," she said, prompting my tears to starts flowing. "You should be hearing from someone after Christmas, and the earliest we can get you here is around February 15, 2009."

I couldn't stop thanking her. My wife and I were in tears and celebrating at the same time after hanging up.

Thriving Through Culture Shock

Now, let's go back to something that was mentioned earlier. The DREAM. Remember, the job offer from the Soccer giants? The soccer coach position was ready in January, but they would get me there in February. Yeah. That DREAM. Remember it was on cobblestone streets in Europe. Do you get it?

I retired January 31, giving the speech about My Three Kings and two weeks later, all seven of us were on our way to the Frankfurt Airport with 12 bags.

Culture shock set in. I left the Marines for the Army, left the coastlines for Europe and departed active duty as a leader to civilian life as a low-ranked "worker-bee."

We immediately found an amazing church in Wiesbaden and they had me over to their house Bible study to share all of what had transpired.

I directed everyone to Matthew 14 where this most amazing Bible story seemed to line up with our current situation. I was all about Jesus. He could do anything for us. Peter was very strong and had a talkative personality also and would proclaim that Jesus was the Messiah with boldness.

When Jesus asked Peter to get out of the boat and walk on water, he did it. He walked on the water, until he realized what he was doing, then his faith weakened, and he began to sink. When Jesus asked me to get out of my comfortable "lifeboat" staring at the retirement storm, I too obeyed and stepped out in faith.

In Fact, my wife and I, we both stepped out in faith. The walk was amazing. The storm had nothing on us.

Yet, whenever we started taking our eyes off of Jesus and looked into the storm we began to sink into the sea.

That phone call from Germany came at the perfect time. When the waves were raging all around, and I started drowning, just then the Lord showed up in a mighty way.

I find it so interesting that for many on a journey with our Savior, we end up sinking to the lowest point and then we're rescued as we surrender all the Him. These challenging experiences are divine lessons from the Lord, designed as building blocks for our faith and empowering us towards achieving the work on earth He's called us to.

And the apostles said unto the Lord, Increase our faith. And the Lord said, If ye had faith as a grain of mustard seed, ye might say unto this sycamine tree, Be thou plucked up by the root, and be thou planted in the sea; and it should obey you.—Luke 17:5-6

Chapter Thirteen

The Good Daily Bread Reminder in the Garden

Then cometh Jesus with them unto a place called Gethsemane, and saith unto the disciples, Sit ye here, while I go and pray yonder.
[Matthew 26:36]

When stationed in Okinawa in 2005, our church took several folks on an adventure of a lifetime to the land of Israel. Never in my wildest dreams would I ever expect to be a part of something so wonderful. As you recall, I spent many years using "Faith Lessons" for Men's Bible studies showcasing different part of Israel and Europe through Scripture.

My first meal there was absolutely amazing. I will never forget going to the salad bar and filling the my bowl with lettuce, tomatoes, cucumbers and so much more. The colors seemed to be so much more vibrant than ordinary salads in the states and abroad.

Biting into the cherry tomato was quite an experience.

"Did you guys taste that?" I asked with eyes wide open. "That cherry tomato had sugar injected into it. I have never tasted one so sweet in my life. Now I know why God told the Israelites that they would be going to a land filled with milk and honey," I said, believing that everything grown there must have honey in it.

Of course, we hit all of the key biblical sites like the Sea of Galilee, Ceasarea Philippi, the Dead Sea and Jerusalem. The 10-day journey was incredible, but in all the excitement, I forgot to call my wife and kids back on Okinawa. She didn't even know if I made it safely. Finally, six days into the trip, I called my wife because I couldn't get money out of an ATM due to the new card we received.

"Hey sweetheart," I said on a cell phone given to me by the tour guide.

"Where are you?" she said. "Why haven't you called me?"

"So sorry, I don't have time, and this is an overseas call," I said quickly. "Can you give me the pin number for our new ATM card?"

After giving me the number, I told her how sorry I was and that we were having a wonderful time—and then I hung up. So crazy.

Nothing was quite as special as visiting the Garden of Gethsemane. Instead of going to the garden site, with a church on it, our Jewish guide was able to get us into the non-tourist area which was on the right side of the road as you face Jerusalem.

Walking into that garden changed my life. Sitting under one of the old olive trees, I started to pray and think about what agony Jesus must have endured as beads of sweat fell as blood to the earth (Luke 22:44). The stress on Him must have been incomprehensible. He was about to endure one of the most brutal deaths ever experienced in history—spiritually and physically, in order to appease the just wrath of God towards the sin of mankind.

This moment in the garden would also be instrumental in another crucial turning point in our family's life journey.

God's Discipline in Marriage

Fast forward to the year 2011. We were living in Germany. I was retired from the Corps and working as a government service employee in Wiesbaden. Remember what I shared earlier about how this job was perfectly placed in our life following my "walk on the water" in faith.

Well, I confess there were some difficult times in our marriage...stemming largely from my pride. I had a tendency, at times, to put my spouse down in front of the kids, but it never dawned on me how much of an adverse effect this was having in our home life. My loud outbursts at times and defending the kids over their mother was like a cancer slowly killing all of us. God, in His mercy, fixed it.

The teens were getting into trouble. The public school system was destroying their minds. My wife caught on to the lies and deceit. Every time she tried to discipline them; a fight would ensue. Instead of defending her, I would take the side of the kids and essentially overrule her authority. At the time, I didn't see this, but you know Who did see it?—*Almighty God!*

What I am sharing here will reveal once again that we have a powerful God who cares for us. He knew that years later, we would be engaged in one of the greatest battles of our lives for our nation. He needed us to be together, and aligned, as a team years prior to the launch of AND WE KNOW®.

Now no chastening for the present seemeth to be joyous, but grievous: nevertheless afterward it yieldeth the peaceable fruit of righteousness unto them which are exercised thereby.
—Hebrews 12:11

God's Perfect Discipline
The pressure cooker finally exploded one day as we discovered many things had gone awry with our two oldest teens. Instead of working together to solve the problem, my wife and I ended up in another outburst in front of our children. This is difficult to share, but necessary to understand the context in order to see how our Heavenly Father comes to our aid in times of trouble.

We had a wonderful home and outwardly things seemed fine. We took vacations together. We enjoyed our church in Wiesbaden. I was preaching many times a month and leading worship. Our youth group was growing. My wife loved Germany and enjoyed shopping for sure.

But the tension when problems hit was too much to handle. How we were addressing the situation was confusing to our children, disruptive to our family, put question marks on our Christian profession, and was certainly not consistent with God's Word for the conduct of husbands and wives.

After that final outburst we instinctively knew that we could not attend the church again. How it all went down was not easy, and it was quite a humbling experience, which, as always, God works out for good.

Our neighbors came over to visit us after they sensed something was wrong. They had stopped attending our weekly Bible study at our home when they discovered we were having issues.

We called Pastor Ron to let him know that we could not attend anymore. Our weekly men's Bible study was taken away from us. The youth nights on Friday nights ceased. Many of our closest friends tried to reach out to us, but we crept into our "cave" and knew it was all gone.

You might be reading this confused. I know. Unless you have been through such an ordeal, it's hard to understand. About a month later, in my desire to fight for my family, I decided we needed to attend our church. Everyone packed into the car, though reluctantly. My wife only got into the car to appease my begging. When we pulled up across the street, she said, "I will stay in the car, you can go." All five of our kids also said, "Dad, we will stay with mom."

I was so exhausted and tired. There was no fight left in me. Even though I was still fighting to get my way, the Holy Spirit was essentially convicting me at the time. That was the moment I knew—this is God's "chastening". I knew I was essentially "spiritually" grounded.

Four Months of Darkness
Instead of diving into the Word of God and looking for answers, I went into deep depression. Can Christians become depressed? You betcha. Whenever we are living in "sin" it is a living hell. What was my sin?

Likewise, ye husbands, dwell with them according to knowledge, giving honor unto the wife, as unto the weaker vessel, and as being heirs together of the grace of life; that your prayers be not hindered.—1 Peter 3:7

That verse really summed it up for me. I pictured my prayers and felt their weight as if they were hundreds of pounds each. Every time I lifted one up, it would rise about a foot in the air and crash back down to the earth. They were being hindered, because of my sin and disobedience.

Instead of surrendering my will back to the Lord, I dug in and did my part in allowing this same sinful routine to continue day in and day out.

In the middle of the fourth month of darkness, we received a call from our good friends in England who planted a Calvary Chapel there. A year prior, we were members in their church, leading worship and participating in the men's Bible study.

During that video call, we started sharing some of the struggles we had with our teens, and our friends quickly sensed something was not right. I don't know how it escalated, but we started arguing right there in front of the pastor and his wife. It was embarrassing. I felt at such a low point—I was done. No more. Prayers stopped.

The pastor's wife courageously interrupted us and said to my wife, "you know I disagree with my husband on many of his decisions, but I have learned to let him lead. It is not easy, but I know my flesh will cause problems if I argue and try to force my way upon him."

I looked at my wife and said, "See, that is what…" (I was about to use that to force submission upon my wife, but was thankfully spared the embarrassment and sin of such a grievous misapplication of God's Word).

"Hold on a second LT," my friend said. "Don't look at her and say anything. Do you know why she follows my lead and doesn't complain? It is because I love her unconditionally. And she knows it and sees it in my words *and* my behavior."

So, now we can see how the Lord was sending us messages of truth and conviction to wake us up to recommit ourselves to each other—the Lord and His Word so we might be empowered from above to restore things in the home?

Several months went by. Each day, I would go straight to my home office after work and watch videos or play a golf video game, or some other time-wasting activity. On the weekends, I would disappear to play golf or spend time at a cafe with my laptop. I was trying to escape, to put it behind me, to ignore the obvious, but I wasn't fooling anyone—those closest to me, and especially the Lord!

After one of my excursions, I came home to a table set for 8. Were we having guests? My wife was in a great mood and started talking with a smile.

"I met some people at the commissary who are from my hometown," she said. "They will be over in an hour. Can you help me get the table ready?"

After the shock wore off, I remained quiet and started cleaning up the bathroom and front porch. We hadn't seen another outside soul in our home in so long. I had no idea what to expect.

For many of you who have been following our channel for several years, you might recall a guest we had on known as KW. His wife really connected with my wife and we planned to meet for the first time.

It was an amazing time and we connected on many fronts. This was a critical moment for me to have an Okinawa family in my life just when I needed it. The Lord sent them to reinvigorate my desire to teach the Bible again. We were talking about our life experiences and backgrounds. Everything was well with our newfound friends, but for some reason my wife commented that we used to attend a church there. I tapped her leg under the table hoping she would understand that I didn't want to discuss it.

Too late.

"You guys go to church?" he asked with a serious look about him. "I have a lot of questions. Maybe you can help me."

"We don't attend anymore," I said. "I would rather not talk about it."

He looked puzzled and started sharing how he attended a church as a teenager and walked down the aisle to pray to receive Christ. Now, I didn't ask him about his conversion experience, he just kept going on and on.

Upon reflection, I'm amazed how our conversation went from "I would rather not talk about it," to "Sure, I will meet you in my office at 6 a.m., Wednesday for Bible study."

He didn't know it at the time, but KW saved my life. He was not a true believer (yet), but God used him in a powerful way just when He knew I needed correction. This was a critical moment for me to have an Okinawa family in my life just when I needed it. To bring me out of a present darkness and into the marvelous light, to want to teach the Bible again. KW's testimony is an amazing story—perhaps told in another book.

Sure enough, in the dark, cold winter German morning, I pulled up to my office where one lone car was parked. KW got out of his car with two cups of coffee and a smile.

We met every week for three months. One-on-one mentorship is the greatest discipleship experience, in my opinion. We covered many faith subjects: the Church, Holy Spirit, the world, gifts of the Spirit, the Enemy, what to expect when we take our last breath, and more.

He followed up with a visit to his family for a month. I didn't hear from him again for about a year.

During that time, God was working in my life and family. You see, when you start teaching the Word of God, it convicts you. The Lord really impressed upon me the importance and responsibility I have been given to get my family in order.

After some fasting and prayer, our Heavenly Father led me to the story of Abraham and how God commanded him to leave his homeland and be cut off from everything he had known his entire life to simply take a walk of faith. God wanted him to leave it all. Leave the comfort of friends, the familiarity of his city, the connections in business. God wanted Abraham to trust Him no matter what! Even when we can't see the bigger picture, the essence of genuine faith is to surrender it all into His will—knowing "all things work for good…"

In similar fashion, though maybe not as dramatic as what Abraham experienced, God pulled me from everything familiar. I was leaving all that was known for many years but was about to learn that my old way of living had to be completely transformed. I needed to start a new journey to a land that was unfamiliar—a peaceful home, God's way. While I don't want to divulge too many details of my younger years, I will leave you with this—my entire upbringing as a child up to the age of 18 was tense and stressful to say the least. Growing up in a peaceful home was not mine to experience.

So, I brought my family into the living room one night with my Bible and a recently purchased calendar. The looks in their eyes said it all. They had the "whatever" eyes. I imagined them saying to themselves "oh, here we go again. Dad is going to get spiritual on us. What a hypocrite."

"Guys, I wanted to let you know that I am truly sorry for the many years you have had to endure my outbursts in this home. I have not been a very good husband or father," I said with a very weak voice and heavy heart.

My wife chimed in quickly, "LT, you are a good father and you treat me great. What are you talking about?" That's my wife. Always sees the best in me.

"I want to ask you to hold me accountable every day," I said, holding up a calendar. "Every day, I am going to record on this calendar when I've said, 'I love you' to my wife and 'I didn't raise my voice' in the home."

We will celebrate each month that I am successful by treating you all to dinner at one of our favorite restaurants in town or we will do something special of your choosing.

Six months later, one of our teen sons made me upset in the car. When I started to raise my voice, it wouldn't work. I had reprogrammed my mind and body. God had helped me in this area in such a miraculous way.

One year later, my wife and I celebrated by going on our first cruise in the Mediterranean Sea. It was as if we were newlyweds.

Eighteen months later, I was told that there was a job available for a GS-12 position in Public Affairs at a location we wanted for many years near my wife's family. I applied for the position and forgot about it. My sole focus each day was family, working hard and keeping my wife happy. Happy wife, happy life, right?

It just so happened that my wife flew to that area to see her family about three weeks after I applied for the position. She took our daughter with her and upon her return seemed to be the happiest I had ever seen.

The Garden

On one particular night, I needed to wake up early for a television event. For some reason my iPad app alarm clock was not working. This was a bit of a shock as this alarm was usually reliable.

So, I just so happened to remember that my blackberry phone had an alarm on it. Now, I never take that to bed with me. I normally left it downstairs next to my keys.

As I dozed off for a good night's rest, I began to dream, and the world opened up as clear as day. There were olive trees around me. Why am I in the Garden of Gethsemane?

For some reason I started praying "The Lord's Prayer." Our Father, Who art in Heaven, Hallowed be Thy Name. Thy Kingdom come, Thy will be done, on earth as it is in Heaven.

Give us this day are daily Br___," trying to get the word "bread" out, but my mouth was sealed.

"Give us this day our daily Br___," I tried but it wouldn't work.

When I tried to get it out the third time, a hand was placed on my left shoulder. I looked over to see a man wearing what I would call "Bible" clothes, only his face was not visible due to the light shining from his face and body.

"I am your daily BREAD!" He said.

Riiiinggggg. Riiiiingggg. Riiiiingggg. I woke up immediately to a phone ringing. You know, the one I brought to bed with me that always remains downstairs.

"Hello," I said looking at the time. It was very early. "Is this LT?" the man asked.

"Yes, it is," I replied.

"We are calling from (Los Angeles)," they said. (Side note: I will not disclose the detailed location in order to protect all partiesinvolved.)

"(Los Angeles), wait a minute," I said. "That's where my wife is. Is my wife, ok?" I asked in a very loud, desperate tone.

You see, she traveled to that area. I assumed since they told me what city they were in that something must have happened to her.

"Your wife?" he asked in a puzzled tone. "No, we are calling about a job you applied for."

"I applied for a job there?" I asked, trying to remember when that happened.

Let's pause here for a moment. Did you catch it? Remember, nearly two years earlier, all that was going through my mind was promotions, Bible study, my youth group, preaching, leading worship. I was especially in tune with knowing everyday if I would get that next promotion or strike an interview.

Yet, now, after reprogramming my mind through Christ, the most important thing on my mind was my wife.

The test was over. The grounding was done. Now for the rest of the story.

"Can we interview you in 20 minutes?" he asked. "We have a government shutdown and need to get this done quickly so we can hire someone right away."

When they called back, the panel introduced themselves and started with the first question. "Why would you want to come to this area and work for this particular organization?"

I asked them how much time we had to which they replied 30 minutes.

"This will take 15 minutes, but after I am done you won't have to interview anyone else because you will know that I am the one chosen to join and support your team."

I shared my story with them, starting back in 1991 when I met my wife and took them on a journey similar to to what I've shared in this book in many ways.

"That concludes my story. Now do you understand why you need me there?" I asked.

Silence.

"L….T…," I heard my name slowly come out of the chief of staff as she was crying uncontrollably. "I have been doing this for 30 years, and I have never been so touched in my life. That was the greatest story I have ever heard."

She continued to speak words of affirmation through her sobs. I felt the Holy Spirit moving through the interview.

One month later, we were on our way to my new position in my wife's hometown. Prior to leaving, we visited our church in

Wiesbaden. When we walked in, everyone came to love on us and hug us—expressing to us how much we'll all be missed. The entire time we were there, I couldn't stop weeping. The love was more than I could handle. Why would they be so kind to us?

The church had grown from eight people when we joined to now around 150. It was so wonderful to see.

About one year after we moved, KW called me and through tears of joy, let me know that he had given his life to our Heavenly Father. One year later, he landed a government job in our hometown. The man whom God used to bring me back to *reality*, was now our neighbor. His wife would also become a Christian and we later baptized them in the ocean.

You see, the Bible is filled with many who have failed our Savior. One particular person I relate to a great deal is the Peter. He seemed so often to have a lot to say. He was an Apostle of immense passion and was endued with power from above, yet he was one who denied his Lord and Savior three times! One might say he was a quitter—but I'd say these low points in his life (as in mine) were times of testing and trial in order to remind us all that in our weakness we are made strong, through Christ (2 Corinthians 12:10).

Just like the Lord waited for me to say "Daily Bread" three times in the garden, he also asked Peter if He loved Him—three times, remember? (John 21)

Peter denied Christ three times and looked into Jesus' face on the final denial. He ran away weeping (Matthew 26).

All his suffering and failure was divine preparation for Peter to become one of the greatest men of Scripture and a pillar of the newly formed Christian Church.

Peter was grieved because he said unto him the third time, Lovest thou me? And he said unto him, Lord, thou knowest all things; thou knowest that I love thee. Jesus saith unto him, Feed my sheep.
— John 21:17

Chapter Fourteen

Vengence is Only Good From the Lord

Learning to let God deal with those who hurt you.

Repay no one evil for evil. Have regard for good things in the sight of all men. If it is possible, as much as depends on you, live peaceably with all men. Beloved, do not avenge yourselves, but rather give place to wrath; for it is written, "Vengeance is Mine, I will repay," saith the Lord.

[Romans 12:17-19]

In this great awakening period, I see many reminiscing about the good ol' days in the 80's when we rode our bikes in packs, stayed out of the house until dark, played basketball for hours, traded marbles and baseball cards, and wore out our board games for hours with friends.

One cultural aspect of those days that seems to come up a great deal in conversation with other parents is child discipline. Many parents discuss the issue of spanking and its overall effect on the health and spiritual prosperity of our families. We had a special board in our kitchen with two small children bending over their bed with their rear ends up, squinting their eyes as they prepared for their wooden correction. The board said, "Never hit a child in the face, nature provided a better place."

The Bible indicates that there is a rod of correction for those who are disobedient.

Foolishness is bound up in the heart of a child;
The rod of correction will drive it far from him.—Proverbs 25:15

When we see foolishness today, the common position for many Christians who obey the Lord would say "That child needs a

Board of Correction" or "They are spoiled rotten."

I was also taught by my Sunday school teachers and mom that there was no need to get revenge on people that hurt you. The things they did would always come back on them. Most who are not influenced by the Word of God might use the word Karma. I don't aspire to that at all.

The Lord seemed to have a good way of always showing me these lessons rather than simply telling me through teachers. He is a brilliant Master of "Show and Tell."

I loved playing basketball in the base housing area of Albany, GA. We would head to that court after breakfast, head home for a quick peanut butter and jelly sandwich and ride our bike right back to the court for more games.

Everyone knew each other. And if a new kid walked up to the court, we would welcome them immediately. Remember, military children move every 2-3 years, on average. We understood each other well.

On one occasion, I happened to make a long shot to defeat one of my new friends in a game of 21. After the win, he was standing in front of me in a display of anger at his unwelcome loss. While he was yelling at me, my eyes kept blinking rapidly. It took me a few "blinks" to understand that he was actually hitting me.

The other guys got around us and started yelling "Fight, fight, fight."

He kept telling me to fight back. I had not anger towards him and said, "No man, I don't want to hurt you." I walked away, got on my bike and headed to the house, not realizing what my face looked like.

"LT, what happened to you!" my mom yelled in horror staring at my face.

"Oh, nothing mom," I said, not wanting any trouble.

Of course, she found out what happened and knew the young guy's mother. She was very upset and stormed out towards her house.

About an hour later my mom returned and shared the most amazing story.

"The boy's mom was very nice, and very upset when she found out that you didn't hit him," my mom said. "She said 'stay right here ma'am while I take care of this.'"

She explained that the boy came in the living room in fear as his mom asked why he hit LT. He said, "I don't know." She took him in the back, pulled out the board of correction and I believe my mom said "he was screaming and crying" as each wallop met his corrective portion of the body.

That guy was one of my best friends after that day. He learned a lesson, but I also learned that God had my back. I didn't have to hit him to win. The real win was letting our Savior do what He does best with His perfect correction…which, in this case, resulted in the debut of our friendship.

These memorable experiences are crucial pieces of the puzzle of my life that the Lord is putting together for His glory and purposeful direction in my life—and keys to what eventually would lead to the birth of *And We Know*.

The Master Guns Vengeance
One of the greatest things I enjoyed in the Marine Corps was leading Marines as a Master Sergeant. The Master Sergeant is Enlisted (E-8) at the 8th level of E-9. Private is E-1.

When you reach this coveted rank, you are referred to as "TOP." When someone would address you, they would say, "Hey Top LT, I am heading to lunch…you want anything?"

I picked up this rank after 17 years in the Corps while stationed on Okinawa. I was put in charge of the Okinawa Marine newspaper, a weekly publication which reached more than 50,000 Americans on the military bases overseas.

There were different areas of responsibility in that Public Affairs office which required several staff non-commissioned officers to handle them, including community relations, media relations, operations and so much more.

We non-commissioned officers would meet with the Public Affairs Chief in charge of all the enlisted every week to go over our area of responsibility. He was an E-9, Master Gunnery Sergeant, who we referred to as "Master Guns." There were only five of these Master Guns in Marine Corps public affairs around the world. He was a big deal.

My role was the toughest as I had about 15 Marines under me working all over Okinawa. Our workload was intense with multiple deployments all over the Pacific.

One day I announced in the weekly meeting, "I am taking my Marines to the golf course Friday after we complete the newspaper, Master Guns."

"You're doing what?" he asked with a confused look. "You can't just take off like that."

I replied, "Master Guns, we worked all weekend, and the Marines need to get to know each other out of the office. We will be doing this quarterly."

He didn't say no, but I knew he didn't want me around.

One of the major reasons was the fact that I would constantly point out where we could be pushing harder against the battalion who kept using our office for their dirty work. If the battalion needed extra hands to move furniture from the barracks, they would call Master Guns and get half of the support from our Marines.

The Master Guns was also very upset because they tried to bully one of my Marines into not informing the battalion commander about an incident he was involved in. The Colonel called my Marine corporal in as a witness to an event involving himself and others. He was taking the risk of not going home for Christmas and facing restriction in the barracks for two months. I stayed quiet the entire time as that young Marine never asked me for advice leading up to that testimony.

I was responsible for driving him to the colonel on the testimony day. As we pulled up, he said, "Top, why won't you talk to me? I need to know what to do."

"You never asked corporal? Since you are asking, I am telling you to go in there and tell the whole story. Tell the truth."

He was told by all those above me to not tell the entire story or he would not be going home for Christmas. He walked into the Colonel's office very nervous, but he came out excited and relieved. He flew home for Christmas. He also came to Christ through this experience, and the Master Guns was livid.

To add fuel to the fire, word got out that a Muslim I visited in the brig, ended up hearing the gospel and starting reading the Bible.

As I took the Marines on their first golf outing in a scramble format, a young lady in a golf cart drove out to tell me I had a phone call. I had to stop playing and head in for the call.

"Hey Top, this is Master Guns. You need to get back to the office to sign some paperwork for the newspaper," he said knowing that I would be upset. When I left the course and got back to the office, he simply pulled me in and said he didn't like that I took them out there for team building.

He made it a habit to give me the worst tasks, requiring weekends away from my family even though he knew the mental toll it was taking on my life and my family.

Bottom line. He didn't like me, I knew it. He knew I knew it. He tried to make my life miserable and find ways to remove me from my office.

He told me to get on a plane and head out right away, even without proper military orders (which is highly unusual) to Thailand. You remember that story. Well, what I didn't share was what happened when the operation was complete.

The Public Affairs Officer, who was a Lieutenant Colonel, took me on a run near our hotel in Thailand. He brought along a good friend of mine who was a gunnery sergeant, E-7.

We drove out into the middle of nowhere. It felt like a movie, and they were about to kill me, bury me and head back to the hotel. Why do I say this? Because they were awkward, they kept looking around to see if anyone was watching and when it appeared we were alone, the Lieutenant Colonel said, "Ok, we can stop right here."

As we paced with our hands clasped on the back of our heads, trying to catch our breath, the gunny looked at me and said, "I owe you an apology, Master Sergeant."

The Lt. Colonel looked at me and repeated it, "I owe you an apology. You were sent out here overnight with no support by someone who recommended you. That person said you would fail miserably. He had me convinced you were the worst public affairs Marine he had seen in his 29-year career," he said with a very sympathetic tone.

"Now Top, in my long career in this field, I have never seen anyone perform at this level," he exclaimed, while putting his hand on my shoulder.

"Yes Top, you killed it devil," added my gunny friend.

The term "Devil" is a short way of saying "Devil Dog" to a fellow Marine. We Marines received that nickname in Germany after a fierce battle in Belleau Wood.

They both added that the thought was I would go down to Thailand for the operation, fail, come back to Okinawa and face serious consequences. The Master Guns really had it out for me, but it backfired.

Then he said, "Somehow, I will make it up to you Master Sergeant. I am not sure how, but it will come."

Now, there were many opportunities to try to fight back against the Master Guns, but the Lord wouldn't let me follow through on it. The Heavenly Father gave me compassion for him beyond what he deserved.

Four months later, while we were in our weekly meeting with the Lieutenant Colonel, it was announced that an E-8 or E-9 requirement came in for a Public Affairs Chief in South Korea. This would fill a space for a one-month military exercise that everyone loathed because it was in tents on a military base next to the Navy Port.

"Master Sgt. LT can go. He is the only E-8 we have," said the Master Gunnery Sergeant with a grin.

"Now hold on there, Master Guns," said the Lieutenant Colonel. "How long has it been since you have deployed?"

"Umm, uh, uh, sir?" the Master Guns stuttered and started visibly turning red in anger. "Sir, I can't go. I don't deploy."

"Sure, you can," said the Lieutenant Colonel, "I think you need to get rid of the cobwebs and get out for a while. It will be good for you."

The Master Guns was livid and told the boss that he needs to talk to him after the meeting; however, the boss told him he was leaving and no meeting was necessary.

Right after the meeting, the Lieutenant Colonel "boss" pulled me in his office and said, "How did that feel, Top? I told you I would make it up to you."

The entire scene reminded me of the story of Haman in the Bible. In the book of Esther, Haman hated Mordecai and the Jews and was looking for ways to take them out. He actually had gallows readied to hang Mordecai on. Yet God turned it all around and the gallows were used to hang Haman instead.

Though the Master Guns was not hung, he was experiencing what he sought to do to me. The story didn't end on a bad note, however.

Three days after the Master Guns departed for South Korea, he called me. I was in shock.

"How is it going Master Guns?" I asked. "Is everything okay?"

"Hey Top, things aren't so good here," he said after a brief hesitation. "I was wondering if you could help me out." "Of course," I said. "I'm here for you, Master Guns."

He said that he heard about my presentations to the generals during the operations and asked how those were put together. It was absolutely amazing. For the next 30 minutes, I filled his email with many presentations and told him how it was done and how to make a huge impression on everyone.

He thanked me over and over before hanging up, but I believe our Heavenly Father led me to say the following:

"Master Guns, I want to let you know that no one in this office will know that you called me for help," I said. "It is important to me that you understand this. I hold you up in the highest regard and will never embarrass you."

He called me back a week later and sounded like he hit the lottery. He was so overwhelmed with joy and asked me for more guidance on other issues like handling the media.

Two months later, a government service worker in our office who filled the role of Public Affairs deputy director, called me during one of the biggest celebrations (my father-in-law's 73rd birthday) I have had with my wife. He did this because of his anger towards me for many reasons.

When the Master Guns found out, he called that civilian into his office and almost had him fired. Not only did he look out for me, but he asked why I hadn't taken the Marines out for golf.

For the next year prior to my departure, the Master Guns would call me to get advice on many issues. Two years later, he was encouraging me to apply for a Master Guns promotion. I retired and got the job as a civilian in Germany.

Well, when I got the job in Germany, the folks there asked me if I knew the deputy Public Affairs officer in Okinawa.

I asked why.

"We ended up picking you over him for the job." Vengeance is MINE saith the LORD.

That, folks, is good vengeance, Amen!

CHAPTER FIFTEEN

LT, the Good Start!

How it All Ties Together

For I know the thoughts that I think toward you, says the LORD, thoughts of peace and not of evil, to give you a future and a hope. Then you will call upon Me and go and pray to Me, and I will listen to you. And you will seek Me and find Me, when you search for Me with all your heart.
[Jeremiah 29:11-13]

One of the most critical areas in life that we've strived to maintain in our family is having dinner together every evening. I see this as a healthy, life-building practice that fewer and fewer families are placing a priority on these days. We see the fracturing of the family and the vilification of traditional family values contributing greatly to the decline of our culture.

Family dinners together is how my sisters and I got to know each other better back in the 70s and 80s. There were times of laughter and tears, joy and pain, jokes and stress, name calling and genuine loving on one another. We would even grow together by cleaning up after dinner.

So, we carried the tradition into our family practice and made it a priority throughout our lives. One of our favorite times was putting our little ones in front of *Little House on the Prairie* while eating. We still remember certain episodes today and end up talking about them when we're together.

The Birth of *And We Know*

I recall sharing with my family at the dinner table one night in

August 2018 about a new influential online intel board I had been following. I told my family how these brave folks were exposing the efforts of nefarious, un-American, anti-God people who were doing all they could to take down President Trump through the lies, deception, and corruption of the mainstream media; the power-mongering of greed politicians and self-professed elites; the socialist ideology of big tech; and more.

As I finished one such story during dinner, my 15-year-old son, Rafael, said, "Dad, you need to record yourself talking like this and get it out to everyone. I know you would get a million followers. I don't understand anything you're talking about, but you make me want to listen."

That statement was the catalyst to what I would call my "Aha moment."

That night, my entire life flashed before me. God had orchestrated my life to be a fighter for our country. My childhood with two dads, living with five sisters, growing up in the Marine Corps, living in the Northeast, meeting my wife who helped me get into Public Affairs, learning how to deal with the media, working in broadcasting, flying into stressful situations, constantly being attacked for my profession of faith in Christ, watching military leaders stand for truth, and so much more.

I departed the government service life to take over a small business that was struggling for survival. We moved from a nice, comfortable home to a tiny 1970's-era worn-down apartment with three bedrooms and one bathroom. The stress was immense, but we were taking steps of faith, knowing God was directing me out of a government role altogether.

Taking the Next Bold Step

One month after my son made that explosive statement, I opened my $450 laptop, attached a cheap microphone to it, opened up my browser to do some research and pressed "record."

The name of the channel was simple. I love the verse Romans 8:28 and had a youth group ministry on Facebook called "All For Good," but the URL I wanted with that name was already taken.

I started researching domains that weren't taken using the verse. I typed in the first three words, "And We Know." To my surprise and excitement, the domain was available, and that day "andweknow.com" was born. I bought that domain immediately and set up a YouTube channel called, **And We Know.**

I then figured out how to make a video intro, render out the video, and upload it to YouTube. At midnight my first podcast was born and alive on the internet. I woke up at 5 a.m. the next morning to start my (other) new business venture, but couldn't stop thinking about *And We Know*.

That night my video had about 50 views and 10 subscribers. The next day, I was at about 100 subscribers.

"I need a t-shirt," I yelled out in excitement. "I can't believe folks are actually watching this."

I found an online design program for customizing shirts. I then found some images that worked together for a possible logo for *And We Know*—a guy holding up a stick; waves of water that I placed on either side of the guy with the staff. I liked the design that was unfolding. I said to myself, "Now that looks a little like President Trump triumphantly leading the American people, emblematic of Moses leading the Israelites out of Egypt, through the parted Red Sea and into the Promised Land."

I will never forget the outpouring of love from folks in the very beginning.

One lady wrote to me saying, "This is amazing. I know you are going to reach millions with this channel."

I chuckled and felt there was no way people would want to listen to the information I was putting out, especially since I was defending the Q board (which many folks had issues with). The one thing that really pulled me in from this "Intel Board" was the fact that there were Bible verses strewn throughout the news reports. It was quite clear to me that not only was this challenging many of my preconceived beliefs of what's happening in the world and prompting me to do my research and wake up to the *Truth*, but it was so refreshing to see how it was all grounded in the Word of God.

I was reminded that this is spiritual warfare that we are in, and that what we would see unfold would be "Biblical." Many YouTube channels started to share this Intel Board, and I could see there was a Christian component to many of them.

My new channel kept growing at a surprising rate. I recall one instance when another influential online channel, *SGT Report*, used one of my videos talking about how the border wall connections I unveiled in my video made a lot of sense. That was a big deal and very exciting for me to see. *SGT Report* was watching our new, fledgling *And We Know*.

"How could a guy with such a huge following pay attention to my video," I told my family.

When the kids would get excited and ask me how it was growing so fast, I told them the same story every time.

"Guys, this didn't happen overnight. It's been in the mind and purpose of God all along. This channel is growing due to the years that the Lord spent molding me and preparing me to be used as a tool for His Kingdom."

What many folks don't know is that I was working from 6:00 am until 4:00 pm in my "day job", then going to a hotel and making my *And We Know* video from 5:00 pm until 1:00 in the morning every day. This was my pace for the first three years of *And We Know*. It was a tough schedule, but I was excited about what was unfolding and compelled to see it through, with God's help! Then, my oldest son, Brandon, came on board and helped me out, making a huge impact.

This new-found "businesstry" was not short of challenges and trials—not unexpected when one commits to stand against tyranny and serve the Lord through truth-seeking.

When the Covid "Plandemic" invaded the world, so many people, businesses, communities were hit hard—and we were no exception. It took months to create the schedule for *And We Know*, but I had to redo the entire schedule in a matter of weeks while trying to create videos with original intros every time. My blood pressure shot up very high and I was exhausted and became overweight.

Through the *And We Know* channel we have seen many people come to know Jesus as their Savior. We have made many new friends around the world. We have had to pray and rely on our Heavenly Father to create for us a whole new *family*.

Our inbox was filled with people sharing how our genuine, humble videos have been used by God to bring them to Christ. We've seen baptisms and photos of thousands of loyal patriots showcasing their *And We Know* gear. It's so rewarding, and a bit overwhelming, to see how many have given their lives to Jesus Christ over these past several years of broadcasting *And We Know*. I am so grateful for the opportunity and privilege of bringing *Truth, Hope, Faith* and *Freedom* to many "truthers", who have now become "family" in so many ways.

We have also had the opportunity to see our children glow, with wide-eyed anticipation as we shared stories about meeting Eric Trump, Kash Patel, Jim Breuer, the Isaacs, Juan O' Savin and so many others. This might seem small to many, but my children didn't really get to see me very often as I was working on a business in the morning and producing videos almost every night. The rewards for that sacrifice are now being realized as we see the fruit of our labor through their eyes.

When we attended the *ReAwaken Tour* in New York in August of 2022, our family had the opportunity and blessing of meeting many *And We Know* fans. Our teens would ask me to walk up to folks wearing the gear and have a conversation with them about the channel. One of the ladies was sharing her love for the channel and paused for a moment as she heard my voice and asked, "Are you LT?" She screamed and hugged me when she realized it was true. The teens still talk about that experience today. They had no idea the impact it was making on people until actually meeting folks who enjoyed the content and were profoundly impacted by our videos.

I always remind them that this channel is God's and he is speaking through us for His purpose. We are now simply enjoying the rewards of belonging to Him.

Epilogue

LT, a Good Testimony!
How it All Ties Together

By faith Joseph, when he died, made mention of the departing of the children of Israel; and gave commandment concerning his bones
[Hebrews 11:22]

Often, I am asked to share my testimony of Christ's calling in my life, and my typical response includes sharing how much I see my life through the story and testimony of Joseph.

His story resonates with me, because God revealed to Joseph that he would have an amazing future, but he had to wait and he would encounter severe persecution along the way as preparation for the revealing of God's plan in his life.

He had a dream that revealed to him a prophetic word that his entire family would one day bow before him. When he shared this dream with his brothers, they were not pleased and instead sold Joseph into slavery in an attempt to rid themselves of him. This heinous act separated him from his father and younger brother and sent him to a foreign land (Egypt).

Like Joseph, I had a sincere desire and sense as a teenager that I would be called and used by God, but the Lord knew so much had to be chiseled away and refined…a lot…before I was ready.

I can relate to the tears and loneliness that must have been devastating to Joseph in those early years of being cut off from his family. Yet, in God's loving providence his experiences were the divine preparation for Joseph to prosper and be elevated as second in charge of the entire known civilized world. Not only did God know how this would work for *good* in Joseph's life, but the Egyptian Pharoah, for whom he was a servant, saw how much Joseph was a man of integrity and trustworthiness and therefore did not hesitate to place him in charge of his affairs—second in command only to himself.

Joseph was cut off from his family for 20 years. He lived most of that time in captivity. As I have often shared with folks, I felt that my time in the "captivity" was when I was in the Marine Corps. The Corps is a great place, don't get me wrong, but the Corps owns you. You can't function without the Corps constantly dictating every facet and detail of your life. For example, when you want to take your family to dinner on the weekend, you better be clean shaven and have the right clothes on or risk being "chewed out" in front of your family for not being compliant to Corps expectations.

Immediately following my retirement, we drove across the country with our five children. I had a short beard, was very relaxed and felt a sense of freedom that had never really experienced while in the Corps. All I ever knew in life was military. I grew up in a military household that was run like the Corps and went straight to boot camp when I was of age, to continue this pattern of compliance that produced a lot of stress.

One day during our cross-country journey, we pulled into a Denny's restaurant and found a round table for the family. While waiting for our meal, the kids started playing with the paper on the straws. As they laughed and made a mess, I joined in on the fun. We were "breaking the rules," but it felt so good, because rules had been imposed on me throughout my entire life.

As we all looked at each other, one of the kids mentioned my smile and how much fun they were having, and it hit me—and tears started flowing. I had to run to the bathroom to hide my emotion. It was the first time I realized what a stranglehold the Corps had on my life and how I imposed this oppression on my own family. It was really okay to have "Fun."

Joseph was interpreting dreams while in the dungeon.

Although he didn't get out at the time he had hoped for, he never ceased worshipping and loving his Heavenly Father. He didn't go into a great depression and give up. He trusted and surrendered himself to God, worked hard, got promoted and moved up in leadership—in favor with God and man (Prov. 3:4; Luke 2:52).

Although I had many life goals and dreams over the years that seemed to provide a future impossible to attain (professional Soccer team coach), I still "pressed towards the mark" knowing that each day with the Lord was wonderful.

Joseph was constantly in danger from those who sought his life—Potiphar's wife and others—but he kept his faith and did not lose heart. Similarly, my marriage and family were attacked from all directions and we experienced severe "dungeon-like" pain and suffering. But no matter what happened we still relied on Jesus Christ.

At the end of it all, Joseph's family was reunited in amazing, unimaginable ways—something only a *good* God can do. Those who hurt him the most were humbled and forgiven. His father and younger brother were returned to him. And his testimony of faithfully trusting in God amidst so many trials is a testimony that should encourage even the most heavy-hearted and worthy of emulation.

Though the world was heading towards a severe famine, Joseph was used by God to restore hope amidst the upcoming ordeal.

Joseph was also blessed with a wife Asenath, daughter of Potiphera, priest of On (Genesis 41:45). I often wonder how wonderful it must have been to enjoy a new culture and a new direction in his life. Again, it is so similar to the way God directed my life. God chose to provide a wife, not at the end of the "captivity" but near the beginning, knowing that she would be the one used to carry me through it all.

I mentioned in Chapter Two about a solid foundation provided through my parents families, yet I did not mention my wonderful in-laws from my wife, Mona. Her father, Rinichi, and mother, Kazuko, showered me with love beyond any measurement one can find on earth. Many times throughout our marriage, Mona's father would say, "Kamisama Jotto (Ka-Me-Saw-Ma Joe-Toe)," which means "God is Good" in Japanese. Mona's mother said she had a dream when she was pregnant that a white dragon flew down and gave her a golden egg, believing that my wife would be blessed even before she was born. The funny thing is that my sisters call me the golden child because they believe I am my mom's favorite, as the only son of six children.

They were both around the age of eleven when World War II broke out on Okinawa. Mona's father said they were lied to about the Americans and told not to eat their food because it was filled with poison. When the American soldiers arrived in their small village in Nago, Okinawa, they offered chocolate to some of the children. Rinichi was so hungry but scared. When he saw one of the American Marines eat the chocolate, he realized the truth and ran with his friends to the Marine and received a token of love amidst a dark world.

Nearly eighty years later, as Rinichi watched his wife Kazuko slowly fading from this earth in 2021, he asked his daughter to stay close to him and his mom Kazuko. They were both told to stay away from Christianity because it was also "poison" in the Japan culture. They witnessed almost every miracle you read in this book and so much more. After my wife poured out the message of salvation one last time, she asked her mom "Do you believe in Jesus?" and her mom said, "Yes." Her mom opened her eyes and talked as if she was no longer ill. Right after Rinichi saw this, he ran to Jesus by telling his daughter "I want what my wife just got."

They both ran to *Jesus Christ*, our *King*. Kazuko experienced a peaceful entrance into glory just a few days later.

God's Plan, My Prayer
I am always amazed when I see the glorious "miracle" of salvation transform an eternal soul to Christ. This is really what the purpose of *And We Know* is all about! The reality of lost sinners that need this salvation is what has fueled my passion for spreading *Truth* and exposing lies; for bringing *Hope* to people yearning for it; to see people's *Faith* revived and florishing; and to see the *Freedoms* we as Americans hold so dear restored! Though

the entire world seems to be heading towards a famine—a spiritual famine—yet God is calling all of us who profess faith in God through Christ, through the power of His Holy Spirit, to bring the message of faith, hope and truth to this lost generation.

Although only a small percent of people relative to the earth's population are tuning in to *And We Know*, it is my never-dying hope and prayer that more and more people will, through our channel, come to know Christ personally. I pray that many would experience this amazing free gift of eternal life that God has planned for all who surrender their lives to Him and trust in the victory of Christ who conquered sin and death on the Cross—through the Resurrection, and where all His people will live with Him in Heaven—FOREVER.

Through it all, I pray that many will come to understand that, in God's loving providence, "All things work together for good to those who love the Lord, to them who are the called according to His purpose (Romans 8:28)."

The Lord's Prayer as found in Matthew 6:9-13 (KJV):

Our Father which art in heaven,
Hallowed be thy name.
Thy kingdom come,
Thy will be done in earth, as it is in heaven.
Give us this day our daily bread.
And forgive us our debts,
As we forgive our debtors.
And lead us not into temptation,
But deliver us from evil:
For thine is the kingdom, and the power,
And the glory, for ever, Amen.

Until next time, this is LT with *And We Know* signing out. Semper Fidelis!